math work

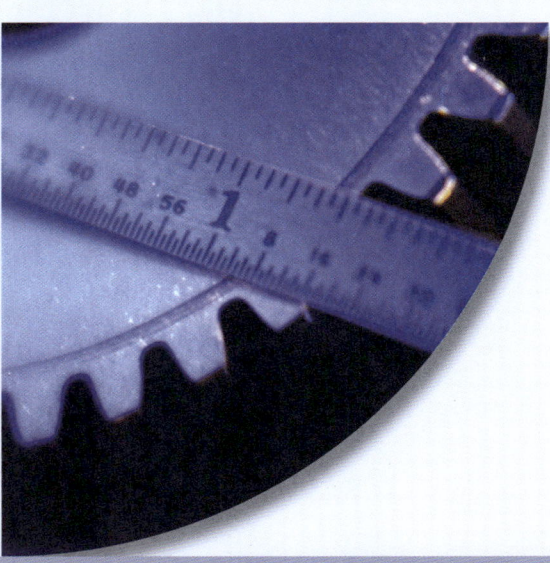

mathwork

By Katherine Trela, Bree Jimenez, and Diane Browder

Editing by Linda Schreiber

Graphic design by Elizabeth Ragsdale

Illustrations adapted from Imagine Symbols by Beverly Potts

Additional illustrations by Gabe Eltaeb

An Attainment Company Publication
© 2008 Attainment Company, Inc. All rights reserved.
Printed in the United States of America
ISBN: 1-57861-638-7

Attainment Company, Inc.

P.O. Box 930160
Verona, Wisconsin 53593-0160 USA
1-800-327-4269
www.AttainmentCompany.com

Table of contents

UNIT 1 Geometry

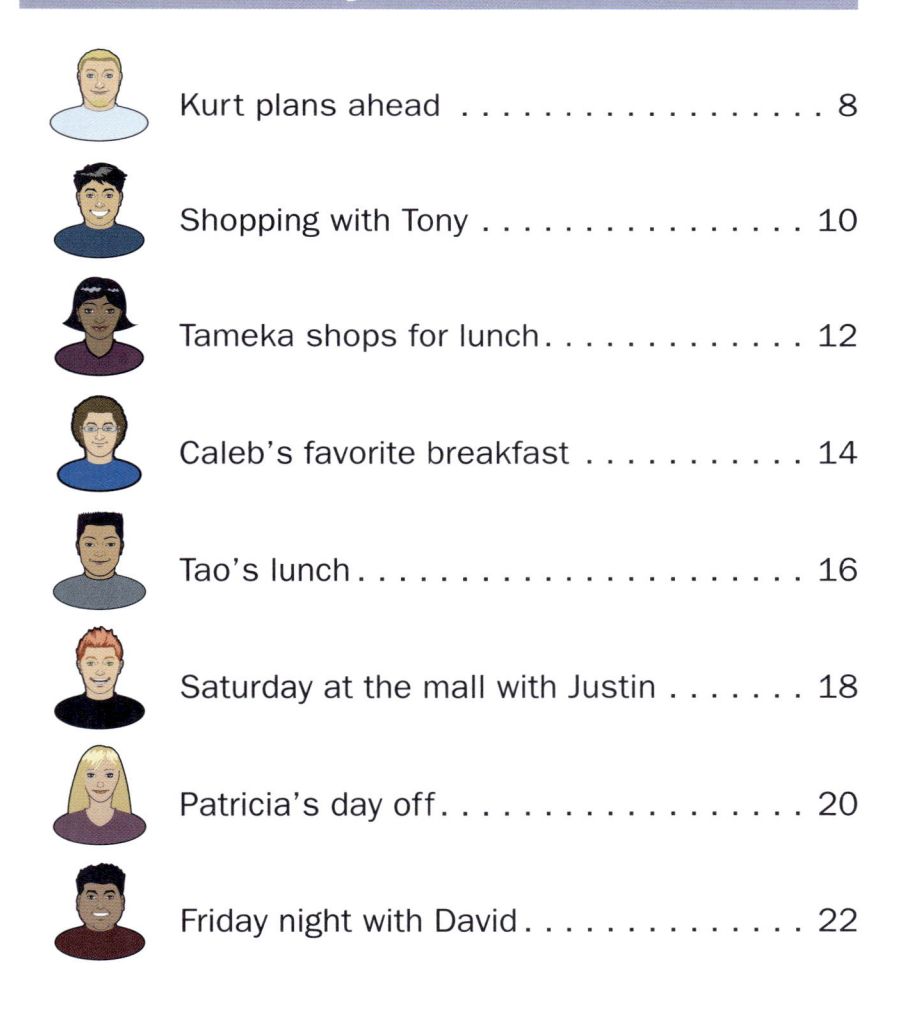

Kurt plans ahead 8

Shopping with Tony 10

Tameka shops for lunch. 12

Caleb's favorite breakfast 14

Tao's lunch 16

Saturday at the mall with Justin 18

Patricia's day off 20

Friday night with David. 22

Dinner and a movie with Meagan 24

Maria at the mall. 26

A place for ice cream. 28

Fancy Feet on the move 30

A new EMT center 32

Arturo's afternoon. 34

After school with James. 36

Liz's busy day 38

Derek buys breakfast 40

3

UNIT 2 Algebra

- Time to shop 44
- At the mall with Jean 46
- A movie for Ben and his brother 48
- Pavel works up an appetite 50
- Gift cards for the movies 52
- Dominick shows a movie 54
- Maria's house party 56
- Movie time 58
- Let's order pizza 60
- Paying for dinner 62
- Ms. Black's class choices 64
- Mr. Williams's students go to class 66
- Mr. Fig's class chooses pizza 68
- Computer class plans future 70
- PE class plans summer fun 72
- Spanish class makes travel plans 74
- Meagan at the food court 76

Table of contents

UNIT 3 Data Analysis

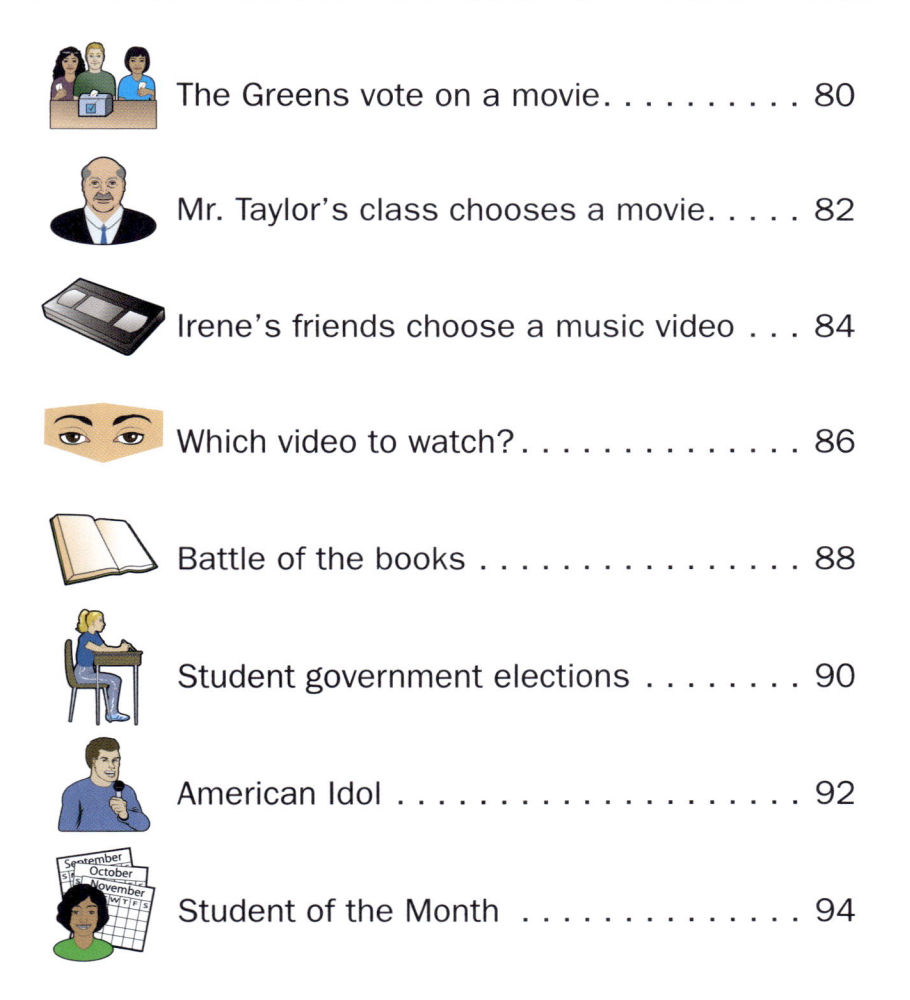

The Greens vote on a movie. 80

Mr. Taylor's class chooses a movie. 82

Irene's friends choose a music video . . . 84

Which video to watch?. 86

Battle of the books 88

Student government elections 90

American Idol 92

Student of the Month 94

A new president 96

Teacher of the Month. 98

Sisters share an apartment. 100

School friends share expenses 102

Pavel's plans 104

Randy's day 106

Favorite classes 108

A class to practice writing 110

The Cruz family rents a movie 112

UNIT 4 Measurement

At the matinee 116

Popcorn at the movies. 118

Casey's family at the drive-in 120

Ordering lunch 122

Out to dinner. 124

Mark at lunch 126

Kate's treat 128

Lunch at the Salad Ranch 130

Shooting hoops. 132

Movie day at the gym. 134

Lee's poster 136

Carla's picture 138

Carter and the rabbits 140

Lee's wall. 142

Carla's wall. 144

Carter's garden. 146

A lunch date 148

UNIT 1 **GEOMETRY**

Kurt plans ahead

Kurt needed to buy food to make breakfast and lunch this week. He needed oranges, ham, and milk. First, he got oranges.

What food did Kurt get next?

Story 1: Kurt plans ahead

WHAT DO WE NEED TO FIND OUT? CHECK THE BOX. ☑

☐ 1. What food did Kurt get next?

☐ 2. What store did Kurt go to next?

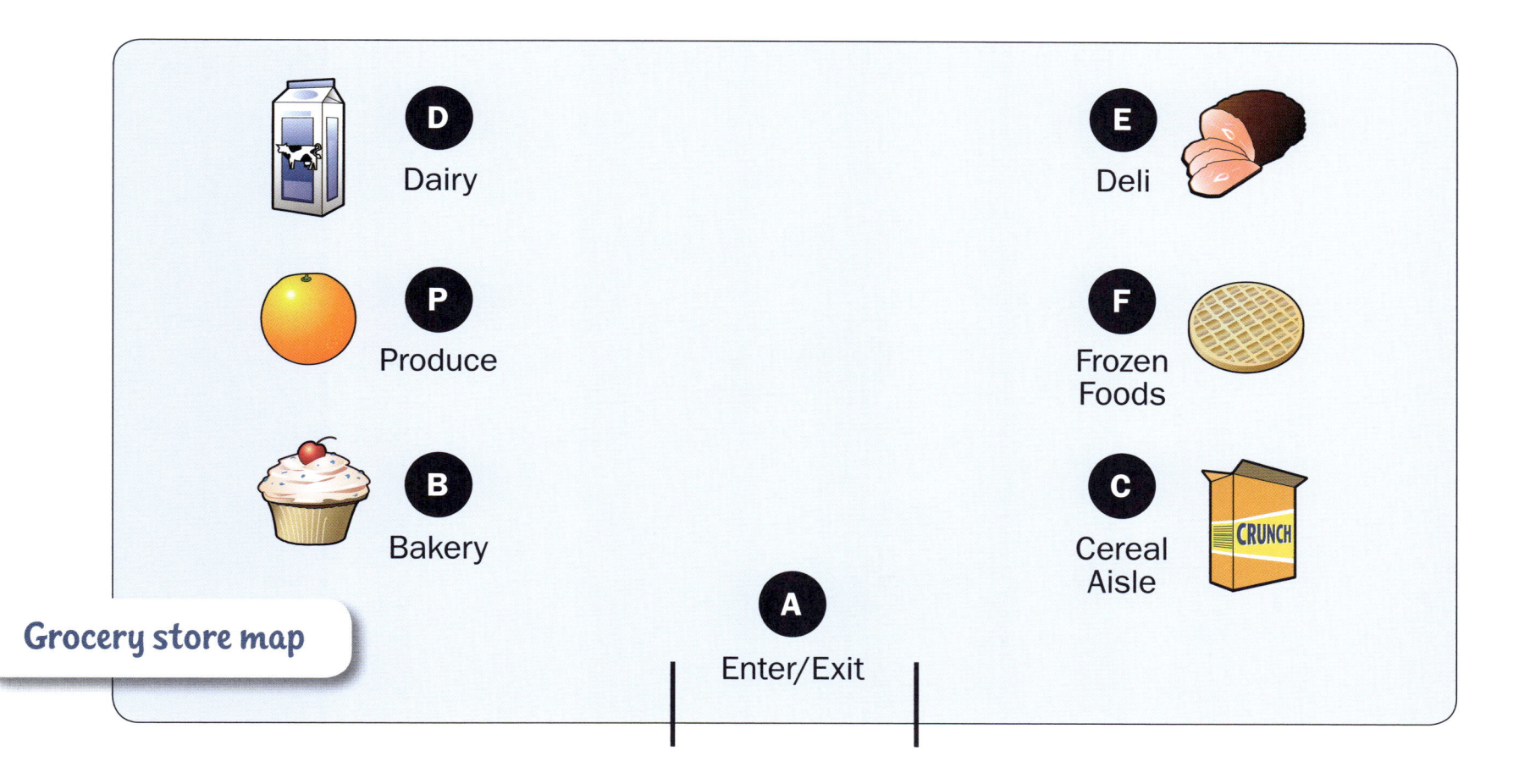

Grocery store map

- **D** Dairy
- **P** Produce
- **B** Bakery
- **E** Deli
- **F** Frozen Foods
- **C** Cereal Aisle
- **A** Enter/Exit

Food: _____

Unit 1: Geometry

9

Shopping with Tony

Tony needed to buy food to make breakfast and lunch this week. He needed grapefruit, frozen waffles, and turkey. First, he got grapefruit. What food did Tony get next?

WHAT DO WE NEED TO FIND OUT? CHECK THE BOX. ☑

☐ 1. How did Tony get to the store?

☐ 2. What food did Tony get next?

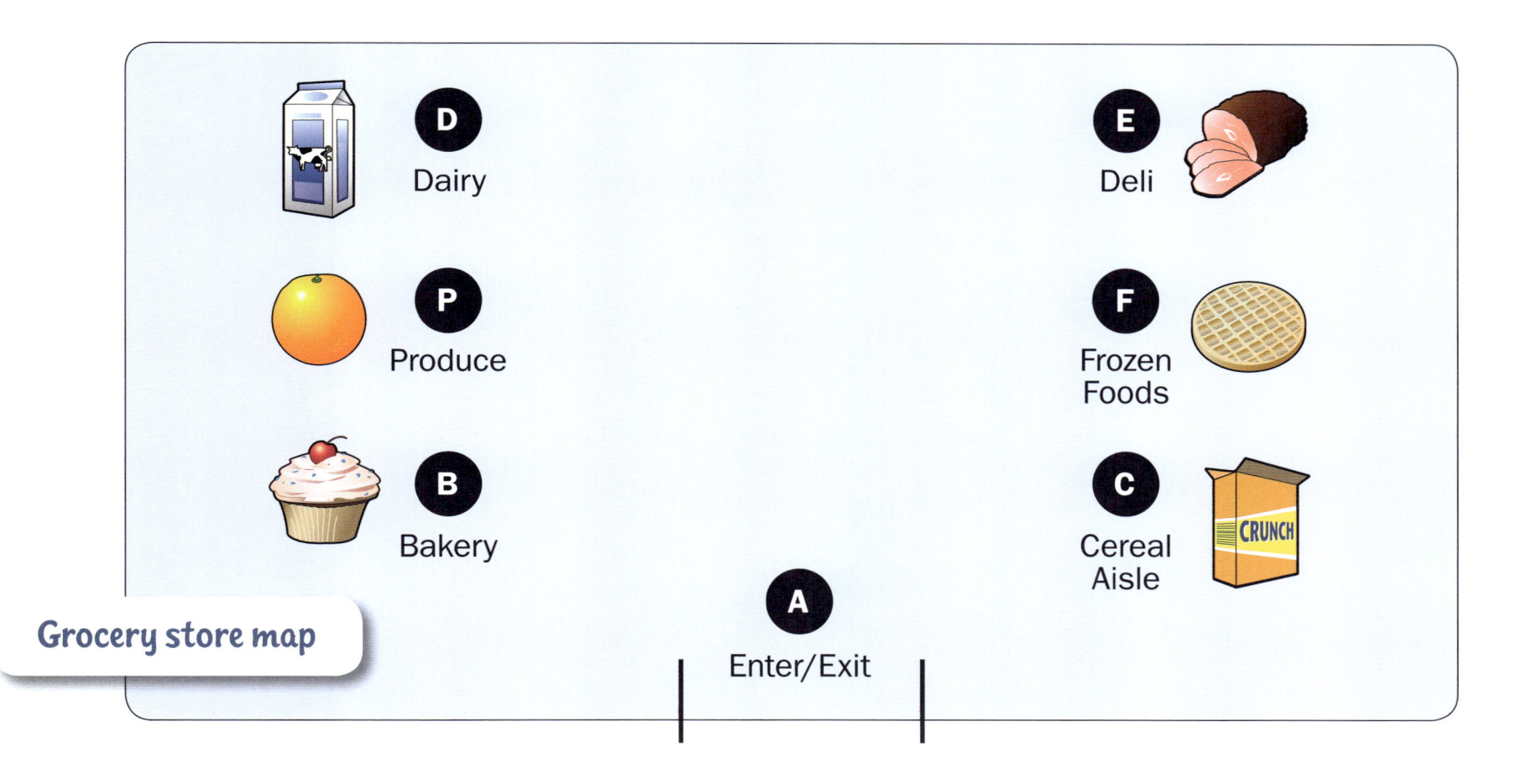

Grocery store map

- **D** Dairy
- **P** Produce
- **B** Bakery
- **E** Deli
- **F** Frozen Foods
- **C** Cereal Aisle
- **A** Enter/Exit

Food: _____

Unit 1: Geometry

11

Tameka shops for lunch

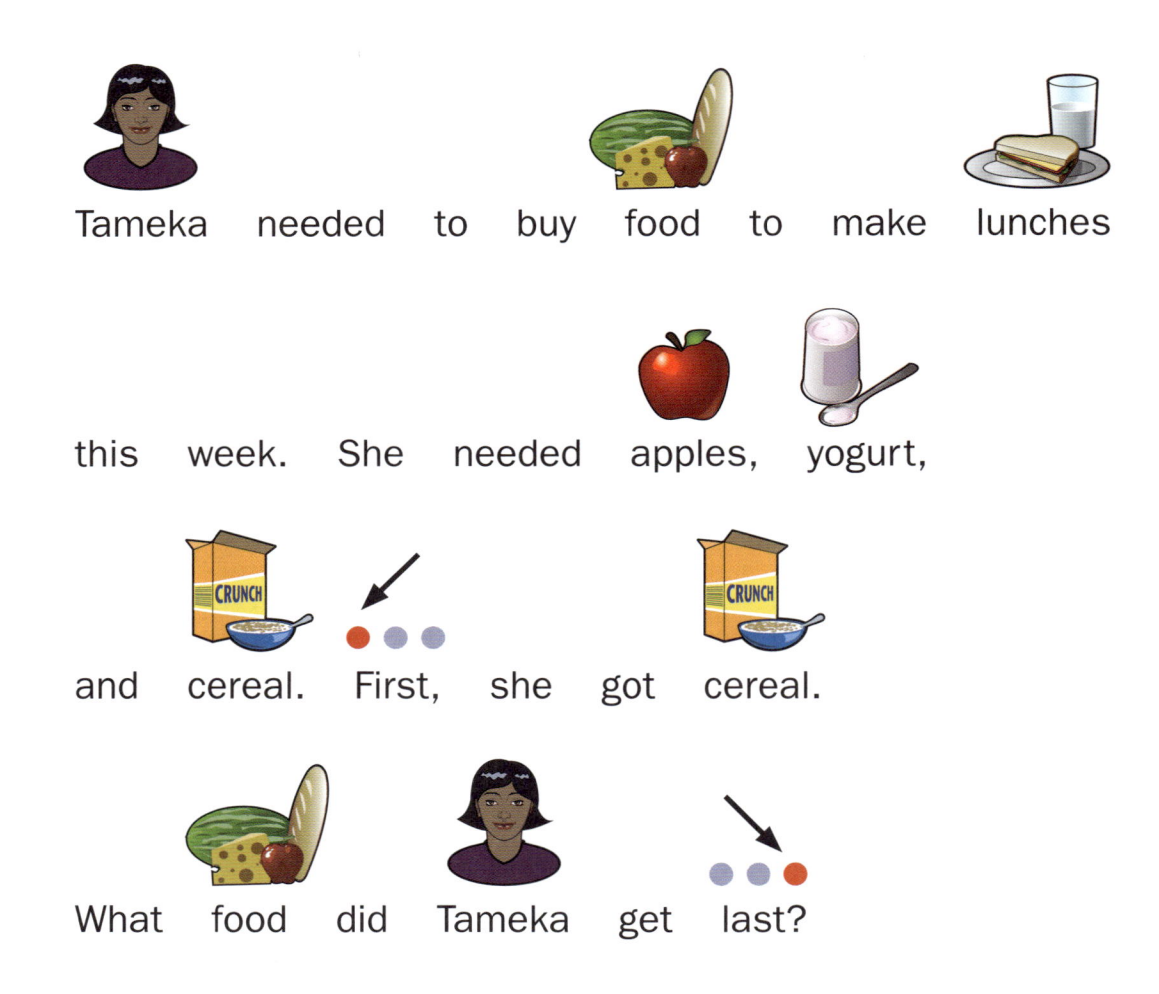

Tameka needed to buy food to make lunches this week. She needed apples, yogurt, and cereal. First, she got cereal. What food did Tameka get last?

WHAT DO WE NEED TO FIND OUT? CHECK THE BOX. ☑

☐ 1. How much money did Tameka spend?

☐ 2. What food did Tameka get last?

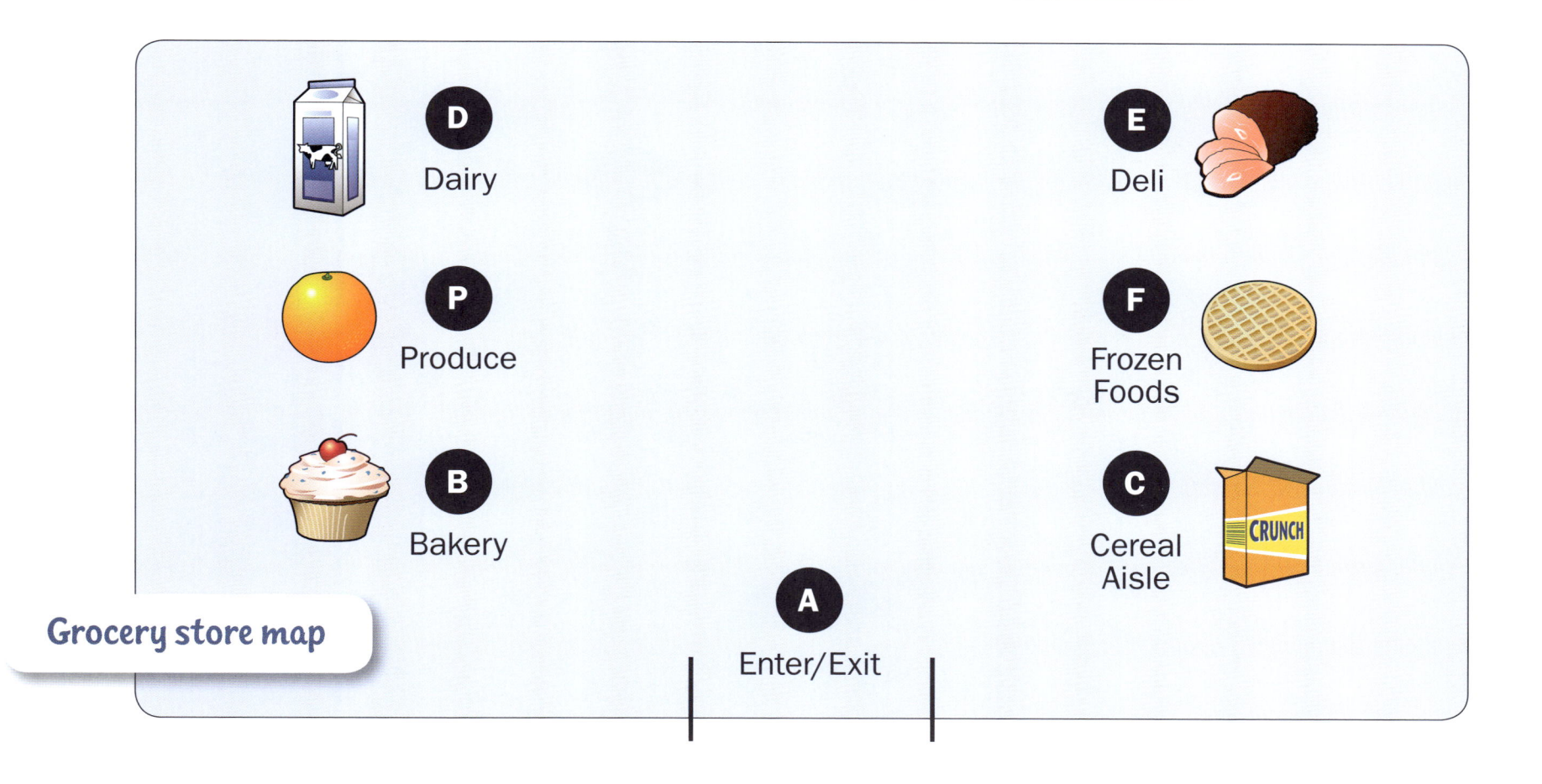

Grocery store map

- D Dairy
- P Produce
- B Bakery
- A Enter/Exit
- E Deli
- F Frozen Foods
- C Cereal Aisle

Food: _____

Unit 1: Geometry

13

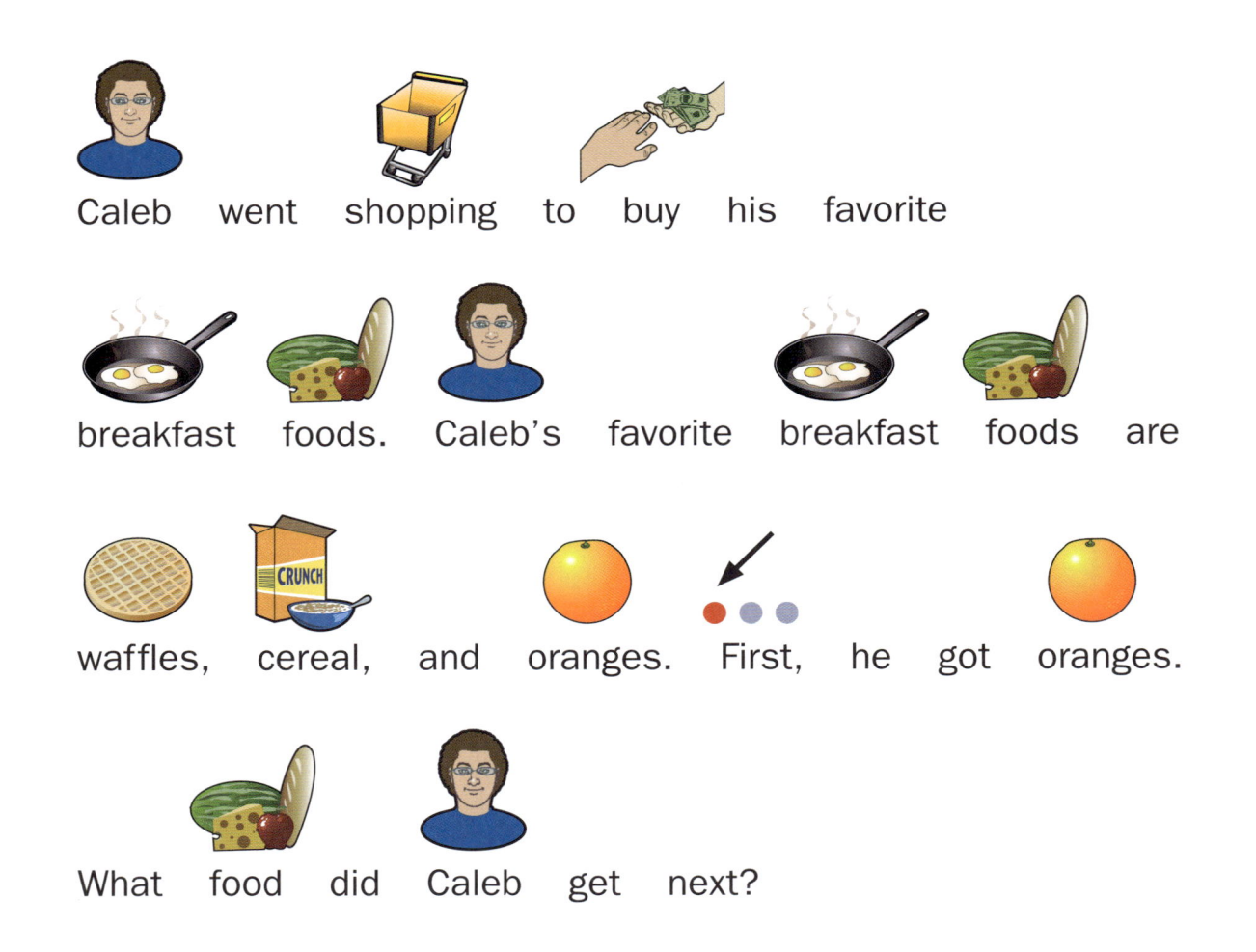

Caleb's favorite breakfast

Caleb went shopping to buy his favorite breakfast foods. Caleb's favorite breakfast foods are waffles, cereal, and oranges. First, he got oranges.

What food did Caleb get next?

WHAT DO WE NEED TO FIND OUT? CHECK THE BOX. ☑

☐ 1. What food did Caleb get next?

☐ 2. What did Caleb eat for breakfast?

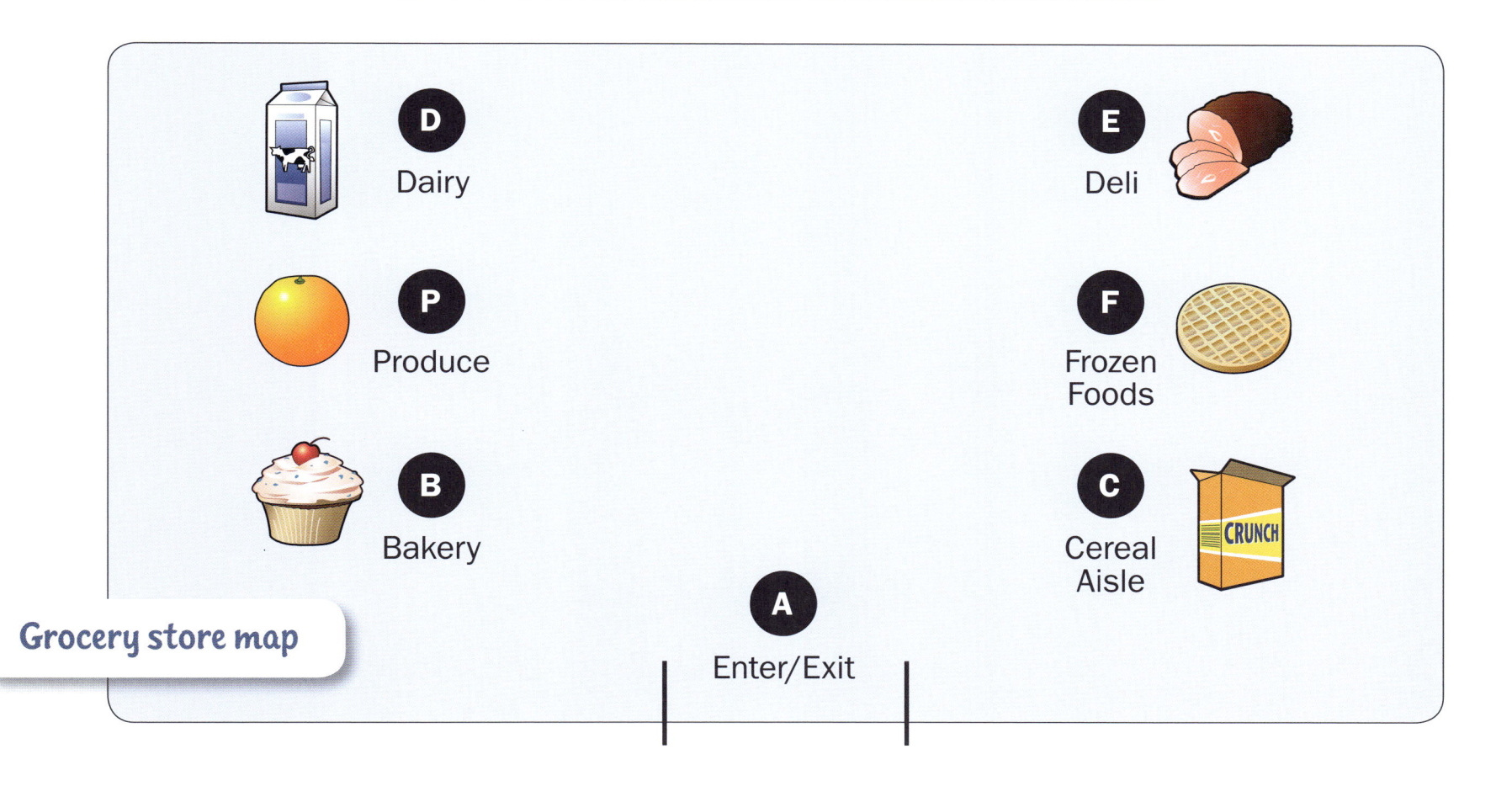

D Dairy

P Produce

B Bakery

E Deli

F Frozen Foods

C Cereal Aisle

A Enter/Exit

Grocery store map

Food: _____

Unit 1: Geometry

15

Tao's lunch

Tao's mom went shopping to make Tao's favorite lunch. His favorite lunch foods are oranges, cheese, and turkey. First, his mom got oranges. What food did Tao's mom get last?

WHAT DO WE NEED TO FIND OUT? CHECK THE BOX. ☑

☐ 1. How much did Tao's mom pay for the food?

☐ 2. What food did Tao's mom get last?

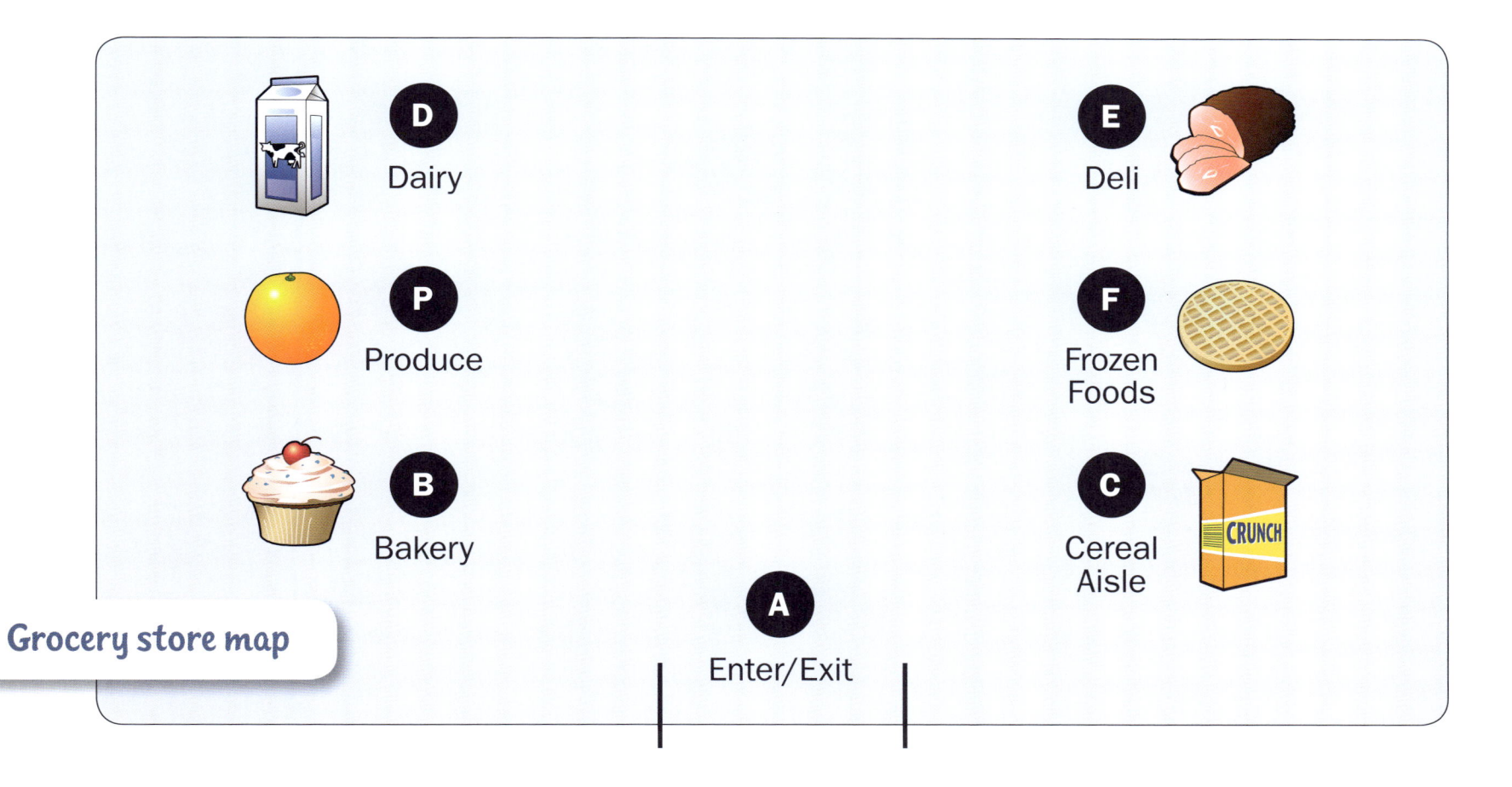

D Dairy

E Deli

P Produce

F Frozen Foods

B Bakery

C Cereal Aisle

A Enter/Exit

Grocery store map

Food: _____

Unit 1: Geometry

17

Saturday at the mall with Justin

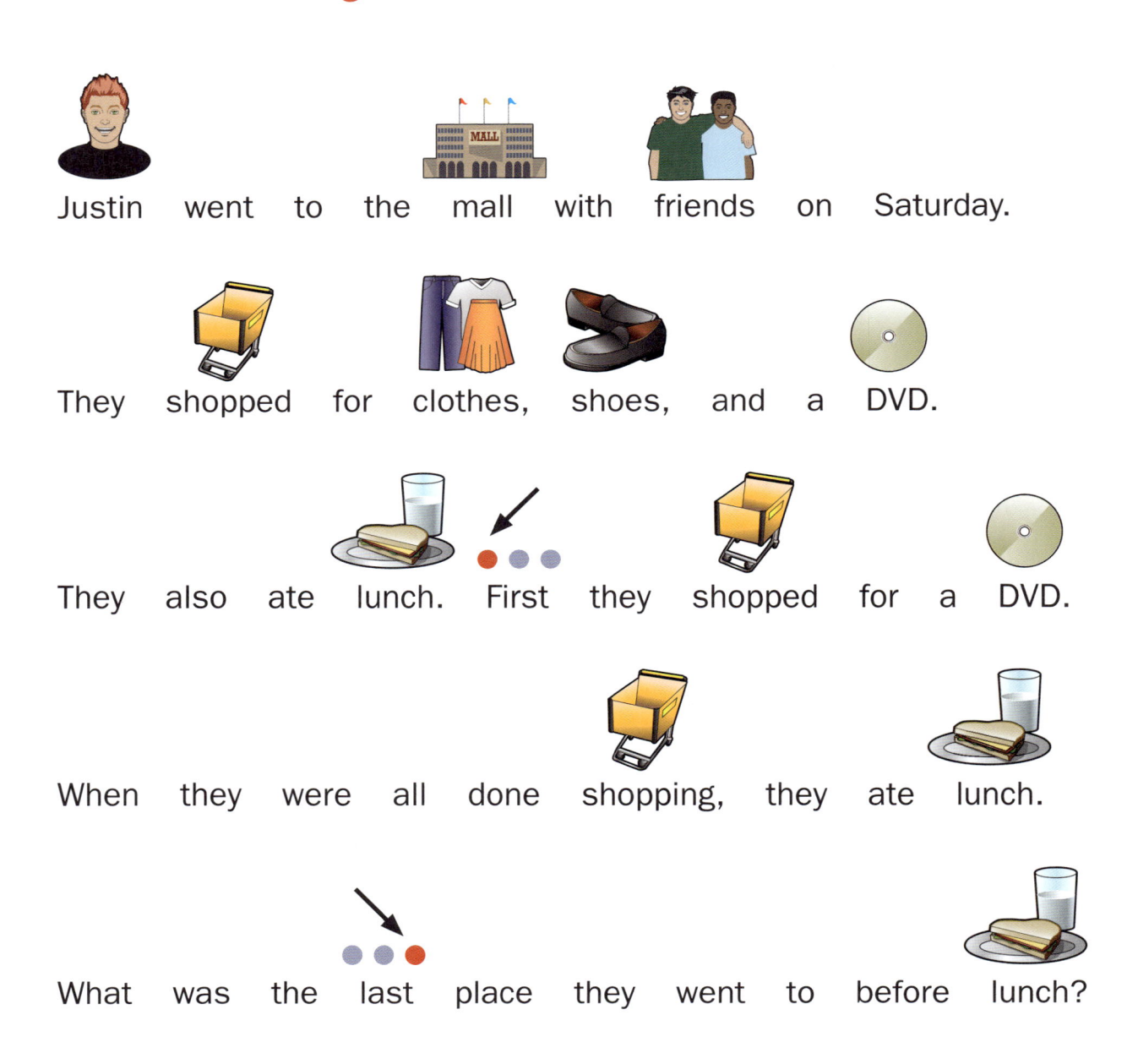

Justin went to the mall with friends on Saturday.

They shopped for clothes, shoes, and a DVD.

They also ate lunch. First they shopped for a DVD.

When they were all done shopping, they ate lunch.

What was the last place they went to before lunch?

WHAT DO WE NEED TO FIND OUT? CHECK THE BOX. ☑

☐ 1. How many hours did they shop?

☐ 2. What was the last place they went to before lunch?

Mall map

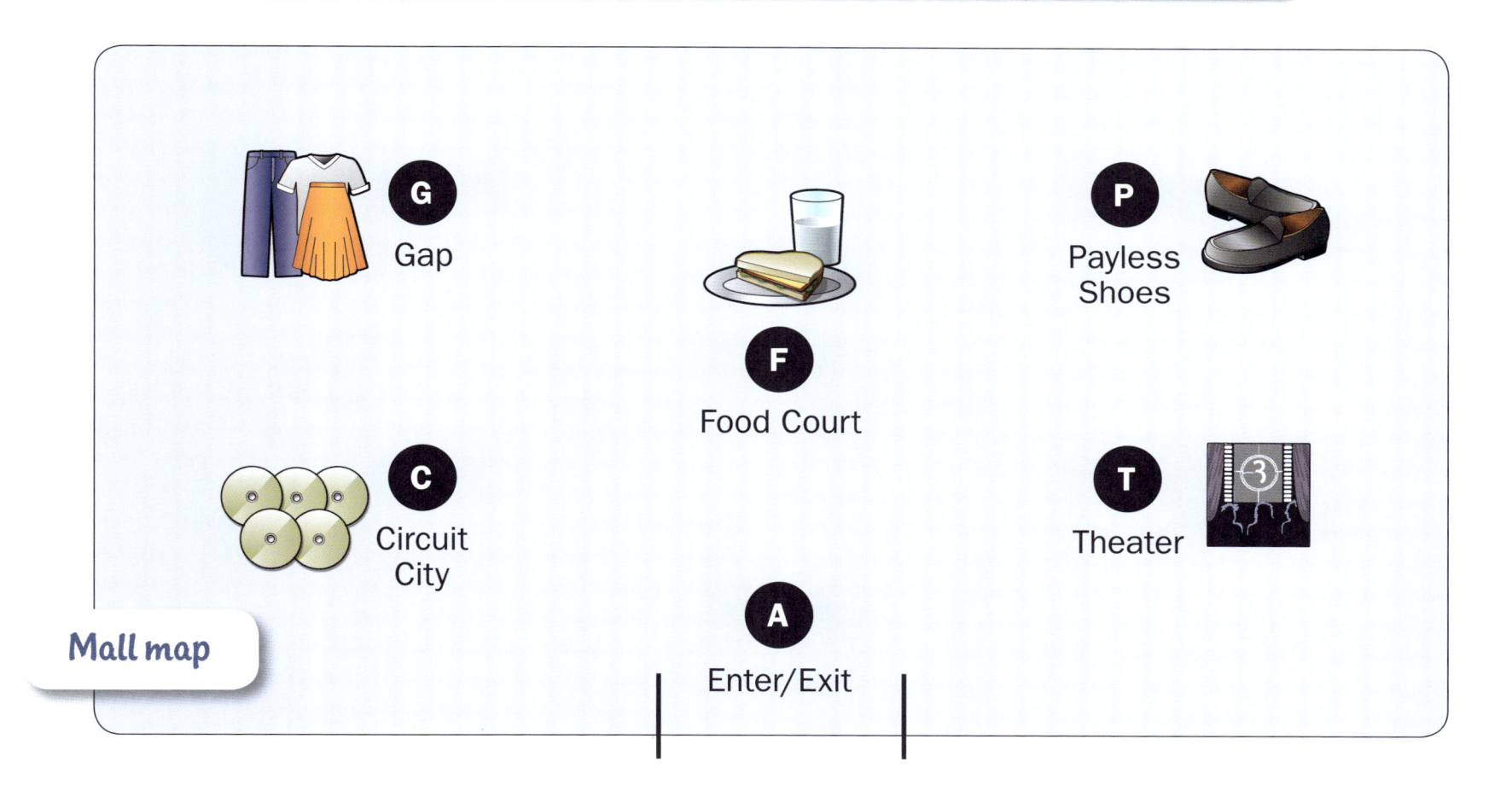

G — Gap
F — Food Court
P — Payless Shoes
C — Circuit City
T — Theater
A — Enter/Exit

Last place: _____

Unit 1: Geometry

Patricia's day off

Patricia went to the mall with friends on her day off.

They shopped for DVDs, clothes, and shoes. They also ate lunch in the mall. First, they shopped for shoes.

When they were all done shopping, they ate lunch.

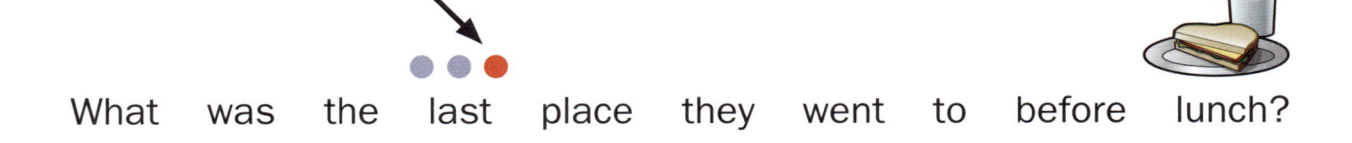

What was the last place they went to before lunch?

WHAT DO WE NEED TO FIND OUT? CHECK THE BOX. ☑

☐ 1. What was the last place they went to before lunch?

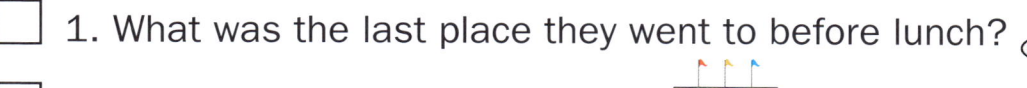

☐ 2. What did they buy at the mall?

G Gap

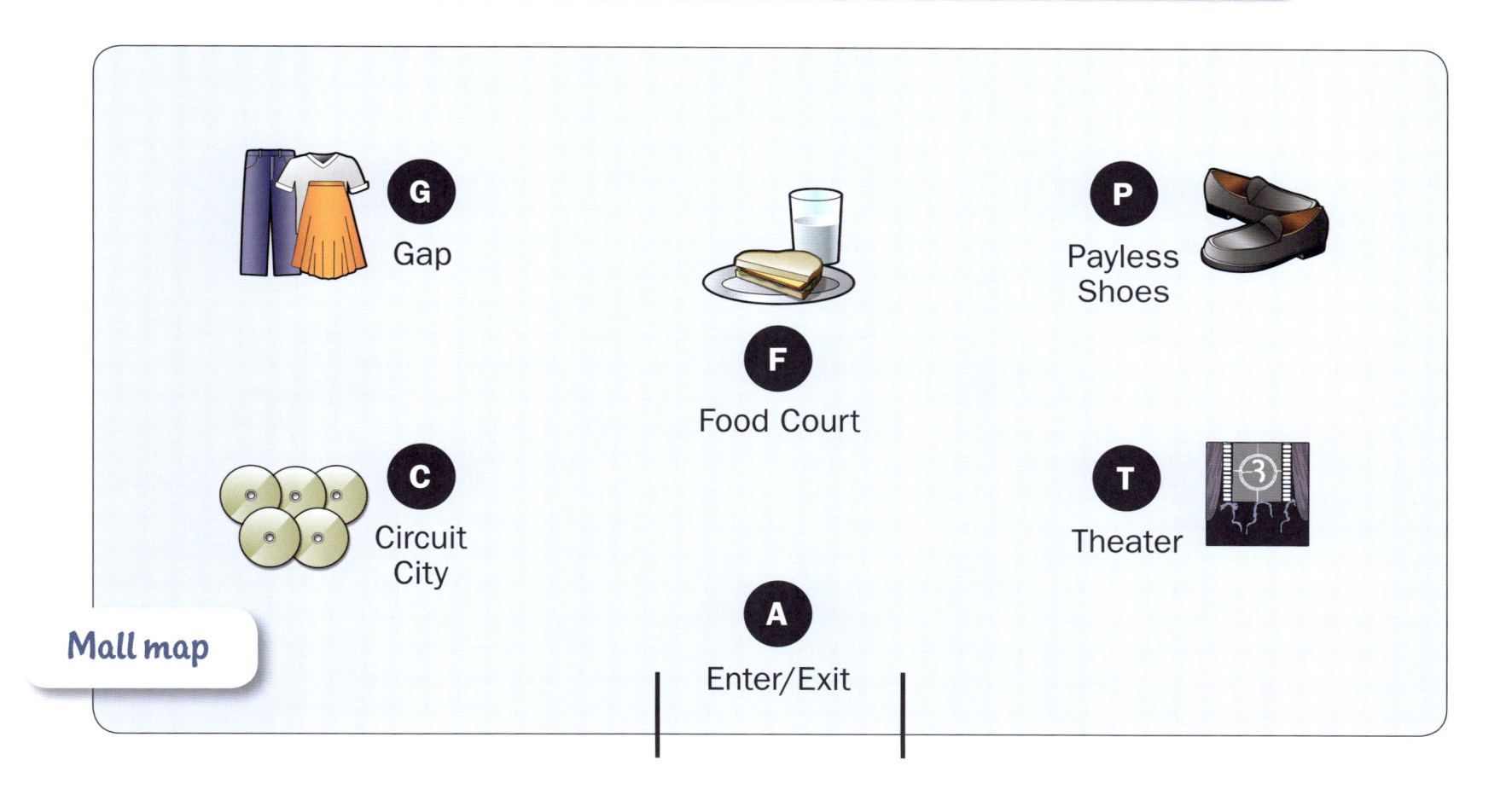

F Food Court

P Payless Shoes

C Circuit City

T Theater

A Enter/Exit

Mall map

Last place: _____

Unit 1: Geometry

Friday night with David

David went to the mall with his sister on Friday night.

They saw a movie, shopped for shoes, and ate dinner.

First, they ate dinner. When they were all done eating and shopping, they saw a movie. What was the last place they went to before they saw the movie?

WHAT DO WE NEED TO FIND OUT? CHECK THE BOX. ☑

☐ 1. What was the last place they went to before they saw the movie?

☐ 2. What did they eat for dinner?

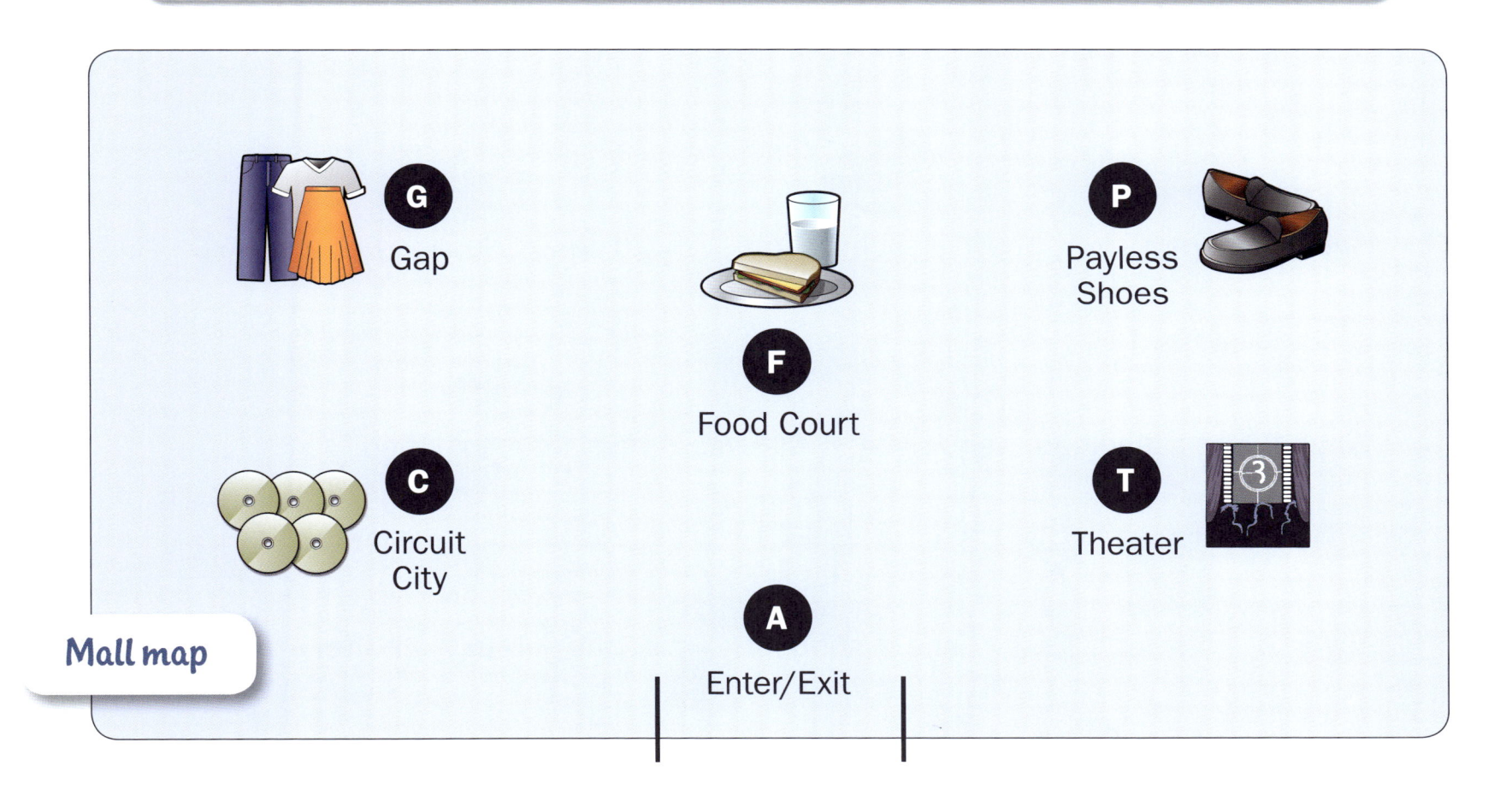

Mall map

- G — Gap
- C — Circuit City
- F — Food Court
- A — Enter/Exit
- P — Payless Shoes
- T — Theater

Last place: _____

Unit 1: Geometry

Dinner and a movie with Meagan

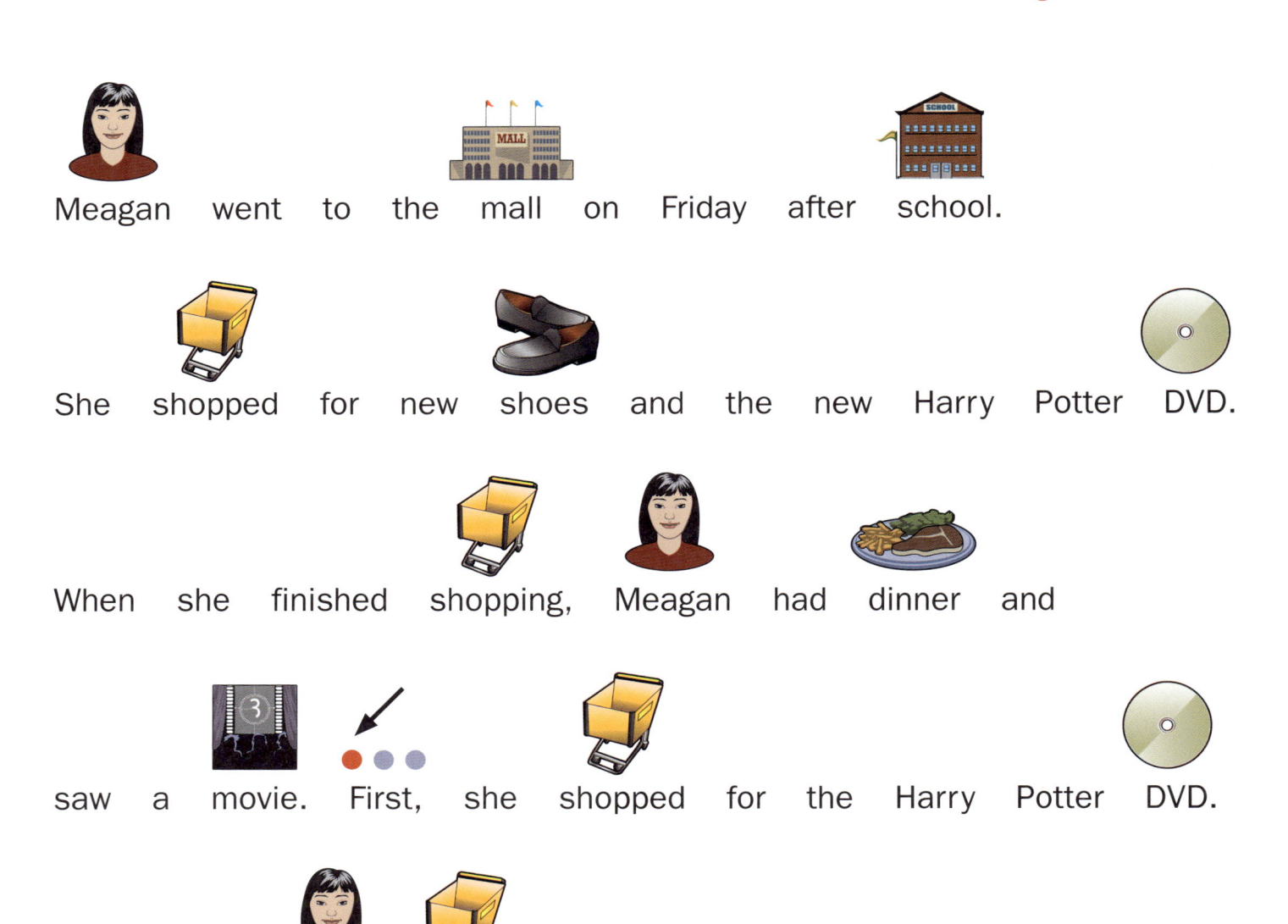

Meagan went to the mall on Friday after school.

She shopped for new shoes and the new Harry Potter DVD.

When she finished shopping, Meagan had dinner and saw a movie. First, she shopped for the Harry Potter DVD.

Where did Meagan shop next?

WHAT DO WE NEED TO FIND OUT? CHECK THE BOX. ☑

☐ 1. What did Meagan eat for dinner?

☐ 2. Where did Meagan shop next?

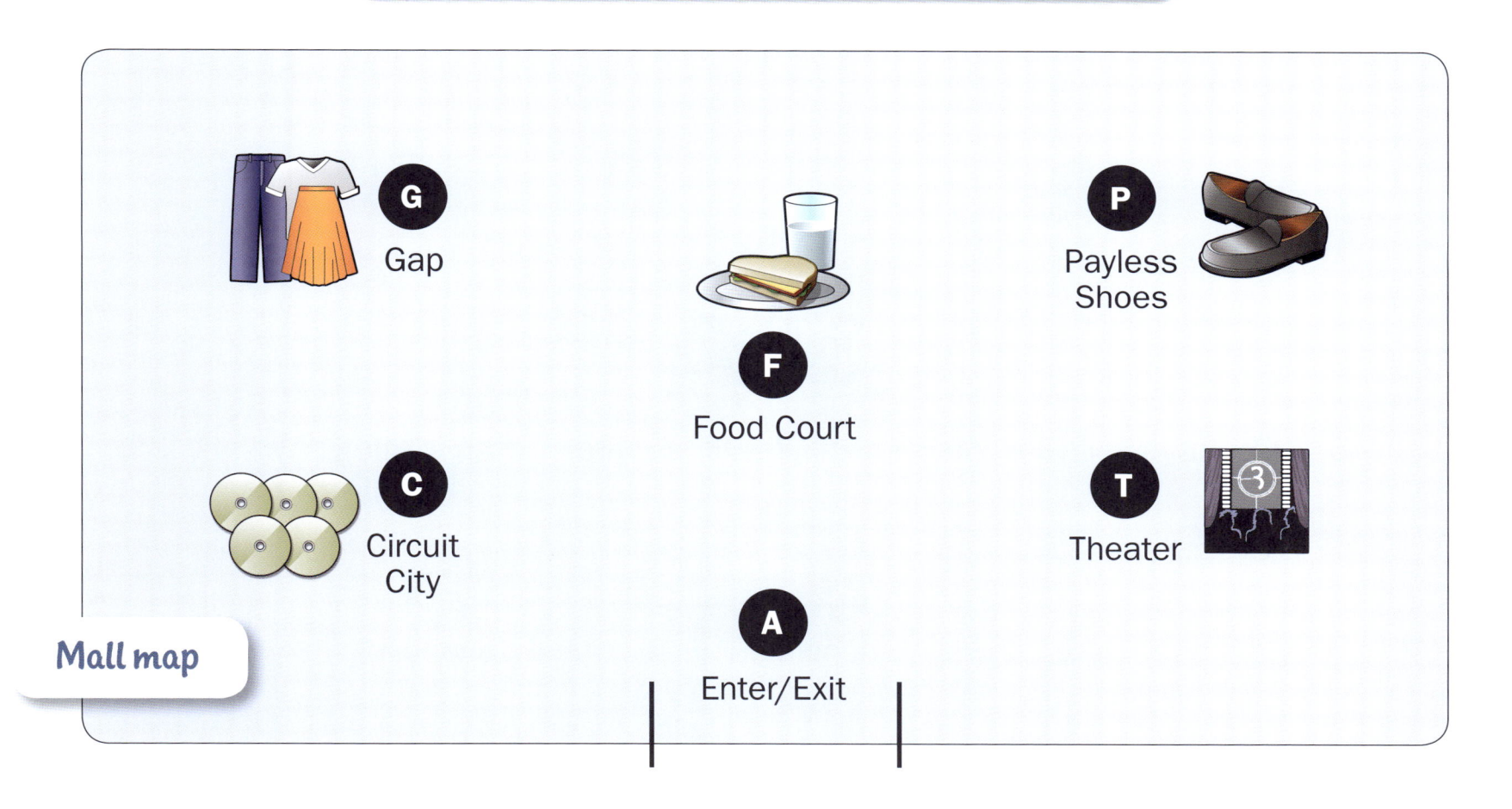

Mall map

- G — Gap
- F — Food Court
- P — Payless Shoes
- C — Circuit City
- T — Theater
- A — Enter/Exit

Next place: _____

Unit 1: Geometry

25

Maria at the mall

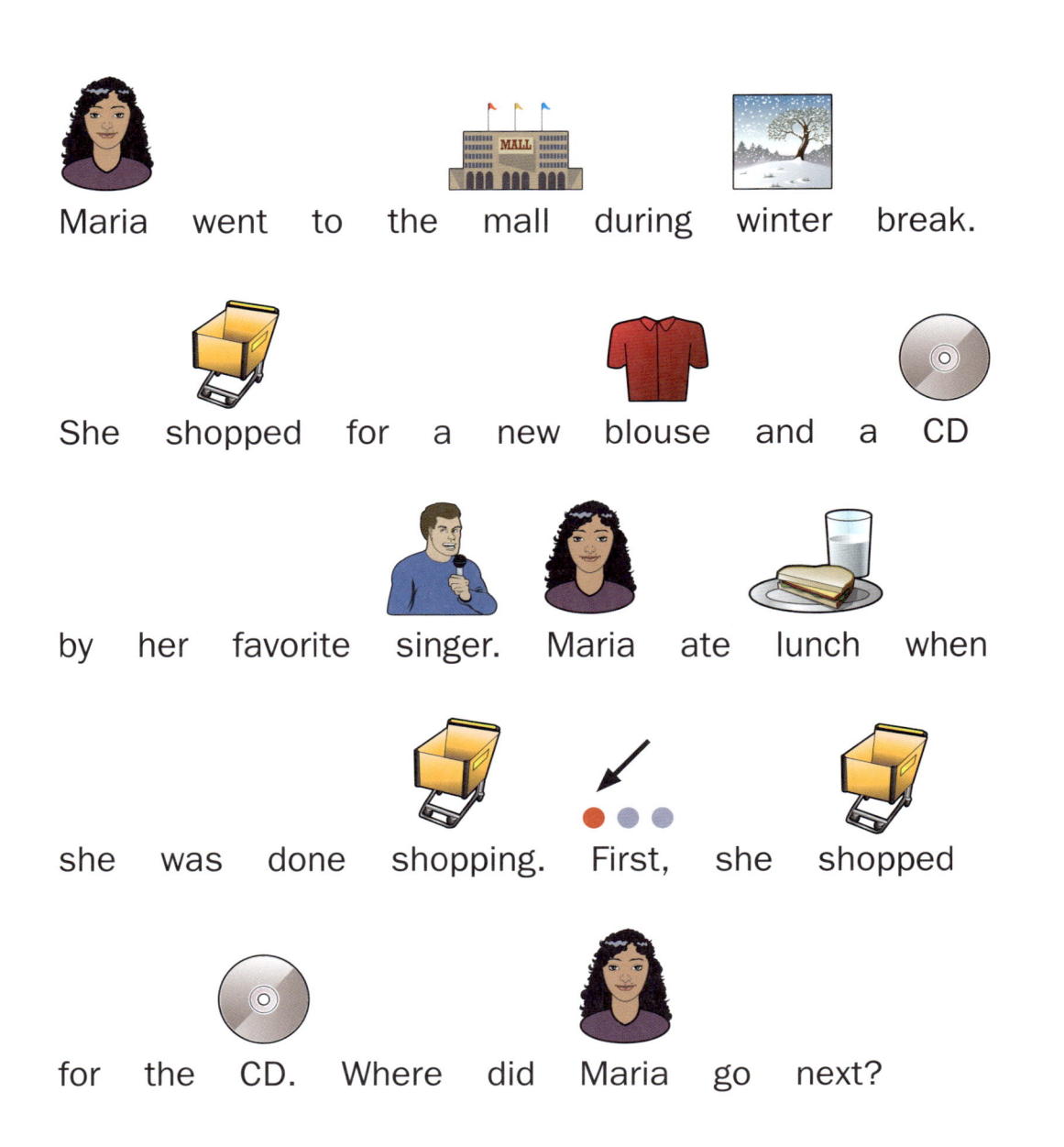

Maria went to the mall during winter break. She shopped for a new blouse and a CD by her favorite singer. Maria ate lunch when she was done shopping. First, she shopped for the CD. Where did Maria go next?

WHAT DO WE NEED TO FIND OUT? CHECK THE BOX. ☑

☐ 1. Why did Maria go to the mall?

☐ 2. Where did Maria go next?

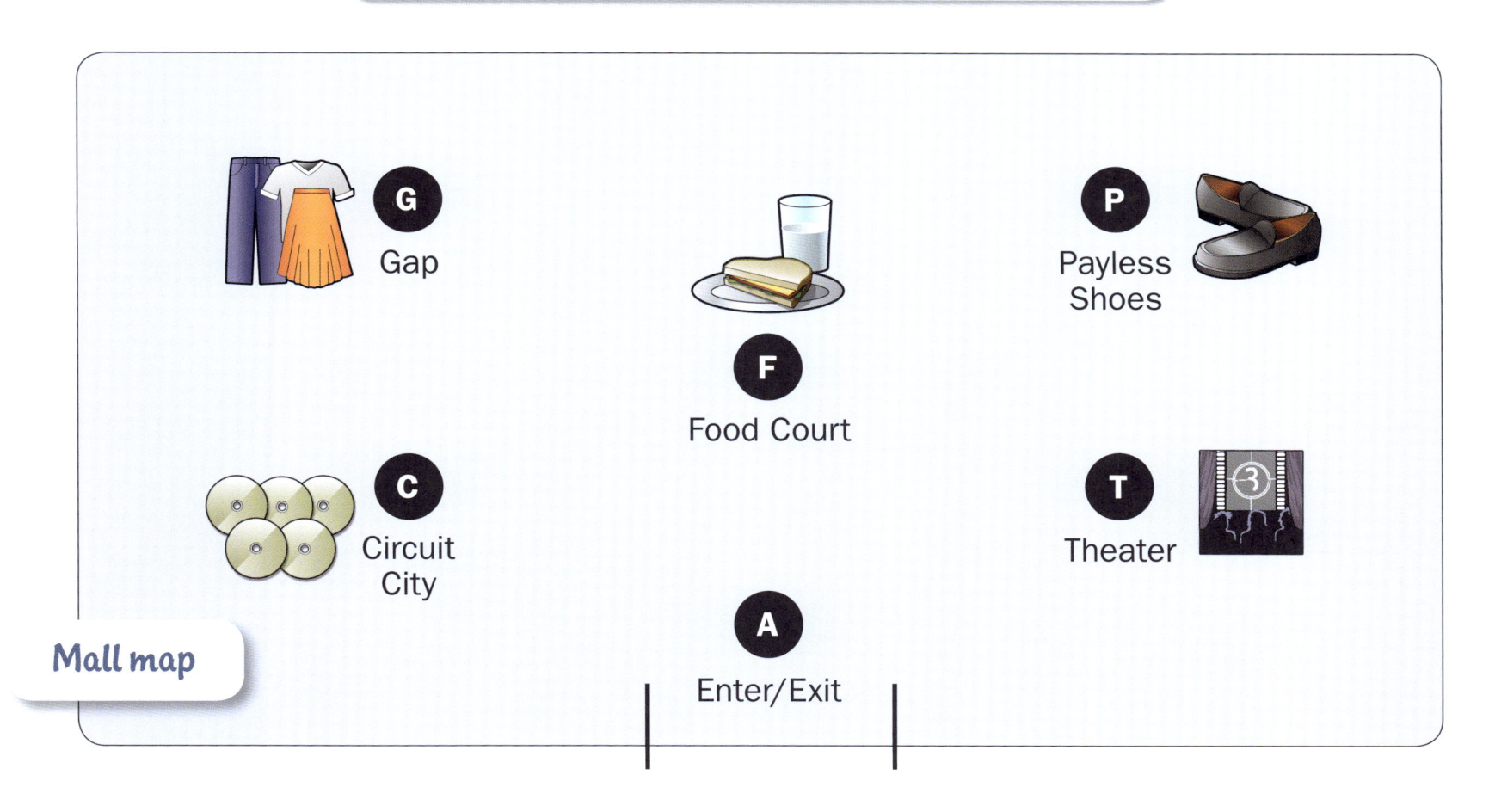

Mall map

- **G** Gap
- **C** Circuit City
- **F** Food Court
- **A** Enter/Exit
- **P** Payless Shoes
- **T** Theater

Next place: _____

Unit 1: Geometry

A place for ice cream

The Softee Ice Cream Company wants to open   a new ice cream shop. They want to be   near a playground. What place on the map   would be good for an ice cream shop?

WHAT DO WE NEED TO FIND OUT? CHECK THE BOX. ☑

☐ 1. How far is the school from the playground?

☐ 2. What place on the map would be good for an ice cream shop?

Street names:

_____

_____

_____

Community map

Unit 1: Geometry

Fancy Feet on the move

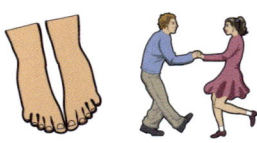

The Fancy Feet Dance Company wants to open a new  dance studio. They want to be near a school and a gym. What place on the map would be good for the  dance studio?

WHAT DO WE NEED TO FIND OUT? CHECK THE BOX. ☑

☐ 1. What place on the map would be good for the dance studio?

☐ 2. How far is the school from the gym?

Street names:

_____

_____

_____

_____

Community map

	A	B	C	D	E	
1	Theater	Grocery Store	School		Police Dept.	
						First St.
2			Park	Gym	Fire Dept.	
						Second St.
3	Library	Diner	Playground	Dentist		
						Third St.
4	Bookstore			Medical Clinic	Hospital	
						Fourth St.

Story Ave. Food Ave. Park Ave. Health Ave. Safety Ave.

Unit 1: Geometry

31

A new EMT center

The  city is looking for a place to  build a new center for the Emergency Medical Technicians  (EMTs). It will have a  garage for the  ambulances. The 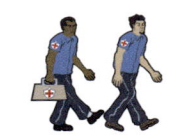 EMT center needs to be near the 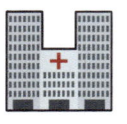 hospital and  fire department. What place on the  map would be good for the  EMT center?

WHAT DO WE NEED TO FIND OUT? CHECK THE BOX. ☑

☐ 1. What place on the map would be good for the EMT center?

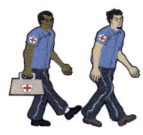

☐ 2. How long will it take to build the EMT center?

Street names:

_____

_____

_____

	A	B	C	D	E	
1	Theater	Grocery Store	School		Police Dept.	
						First St.
2			Park	Gym	Fire Dept.	
						Second St.
3	Library	Diner	Playground	Dentist		
						Third St.
4	Bookstore			Medical Clinic	Hospital	
						Fourth St.
	Story Ave.	Food Ave.	Park Ave.	Health Ave.	Safety Ave.	

Community map

Unit 1: Geometry

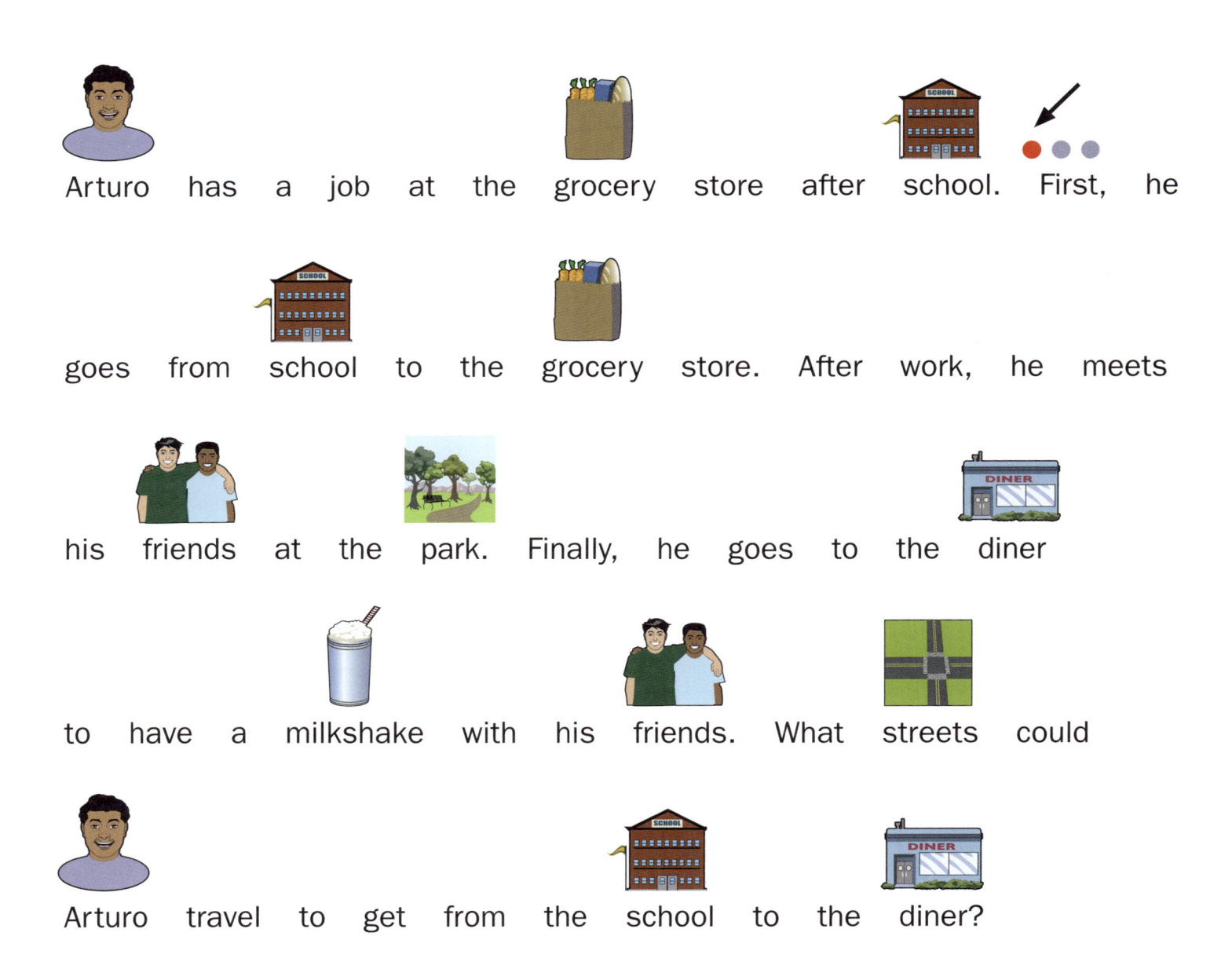

Arturo's afternoon

Arturo has a job at the grocery store after school. First, he goes from school to the grocery store. After work, he meets his friends at the park. Finally, he goes to the diner to have a milkshake with his friends. What streets could Arturo travel to get from the school to the diner?

WHAT DO WE NEED TO FIND OUT? CHECK THE BOX. ☑

☐ 1. How long has Arturo worked at the grocery store?

☐ 2. What streets could Arturo travel to get from the school to the diner?

Street names:

_____

_____

_____

	A	B	C	D	E	
1	Theater	Grocery Store	School		Police Dept.	
						First St.
2			Park	Gym	Fire Dept.	
						Second St.
3	Library	Diner	Playground	Dentist		
						Third St.
4	Bookstore			Medical Clinic	Hospital	
						Fourth St.

Story Ave. | Food Ave. | Park Ave. | Health Ave. | Safety Ave.

Community map

Unit 1: Geometry

35

After school with James

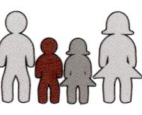

James helps watch his little brother after school.

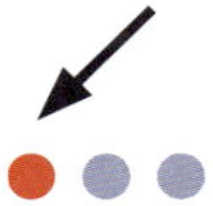

First, he picks up his brother at the school.

Next, they go to the playground. Finally, James 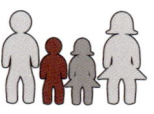   and his brother go to the library. What streets   could they use to get from the school to the library?

Story 15: After school with James

WHAT DO WE NEED TO FIND OUT? CHECK THE BOX. ☑

☐ 1. How old is James's little brother?

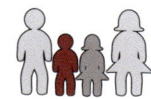

☐ 2. What streets could they use to get from the school to the library?

Street names:

_____

_____

_____

	A	B	C	D	E	
1	Theater	Grocery Store	School		Police Dept.	
						First St.
2			Park	Gym	Fire Dept.	
						Second St.
3	Library	Diner	Playground	Dentist		
						Third St.
4	Bookstore			Medical Clinic	Hospital	
						Fourth St.
	Story Ave.	Food Ave.	Park Ave.	Health Ave.	Safety Ave.	

Community map

Unit 1: Geometry

Liz's busy day

Liz volunteers at the  theater.  Today she has a  dentist appointment. She must return a  book to the  library before she goes to the  theater, too. What  streets could  Liz use to get from the  dentist to the  theater?

WHAT DO WE NEED TO FIND OUT? CHECK THE BOX. ☑

☐ 1. What streets could Liz use to get from the dentist to the theater?

☐ 2. What book did Liz read?

Street names:

_____

_____

_____

_____

Community map

Unit 1: Geometry

Derek buys breakfast

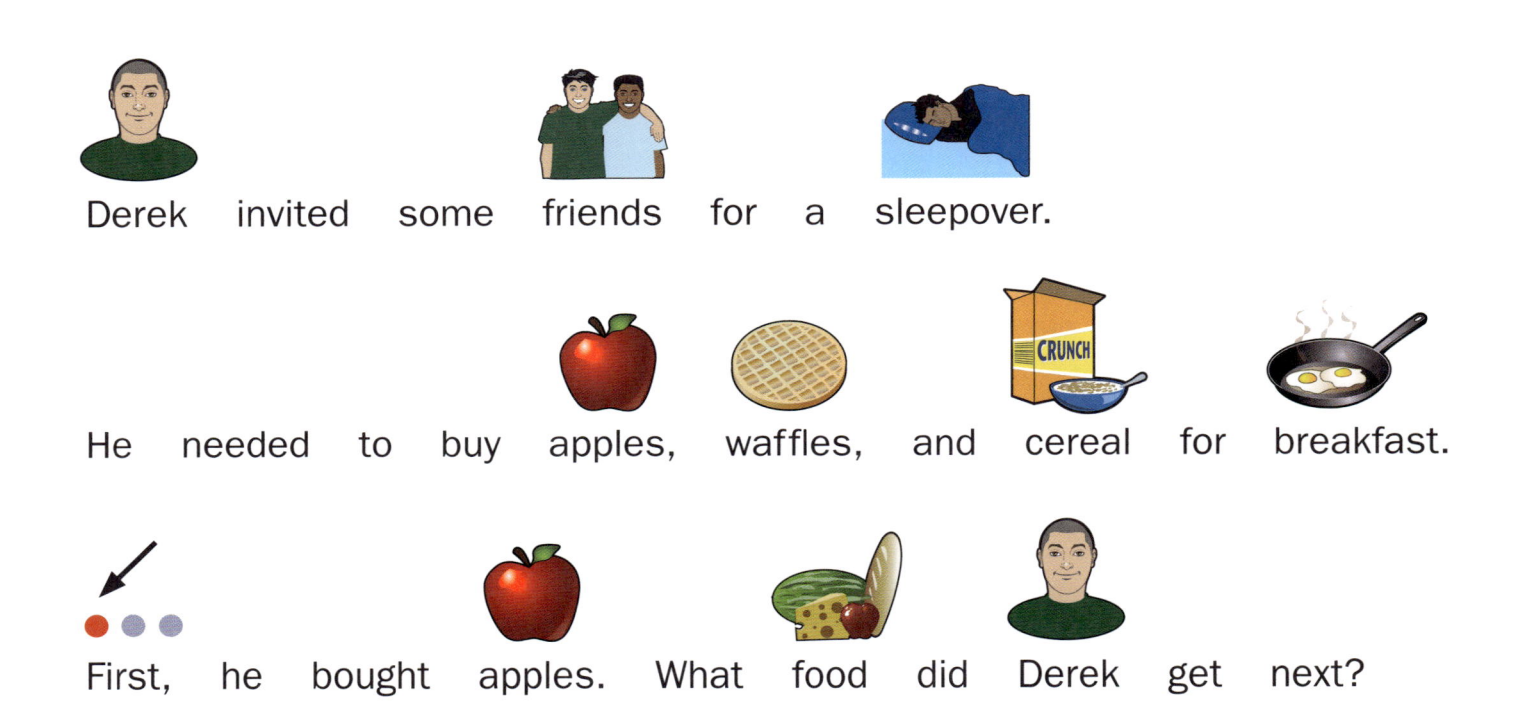

Derek invited some friends for a sleepover.

He needed to buy apples, waffles, and cereal for breakfast.

First, he bought apples. What food did Derek get next?

WHAT DO WE NEED TO FIND OUT? CHECK THE BOX. ☑

☐ 1. How many dollars did Derek give the cashier?

☐ 2. What food did Derek get next?

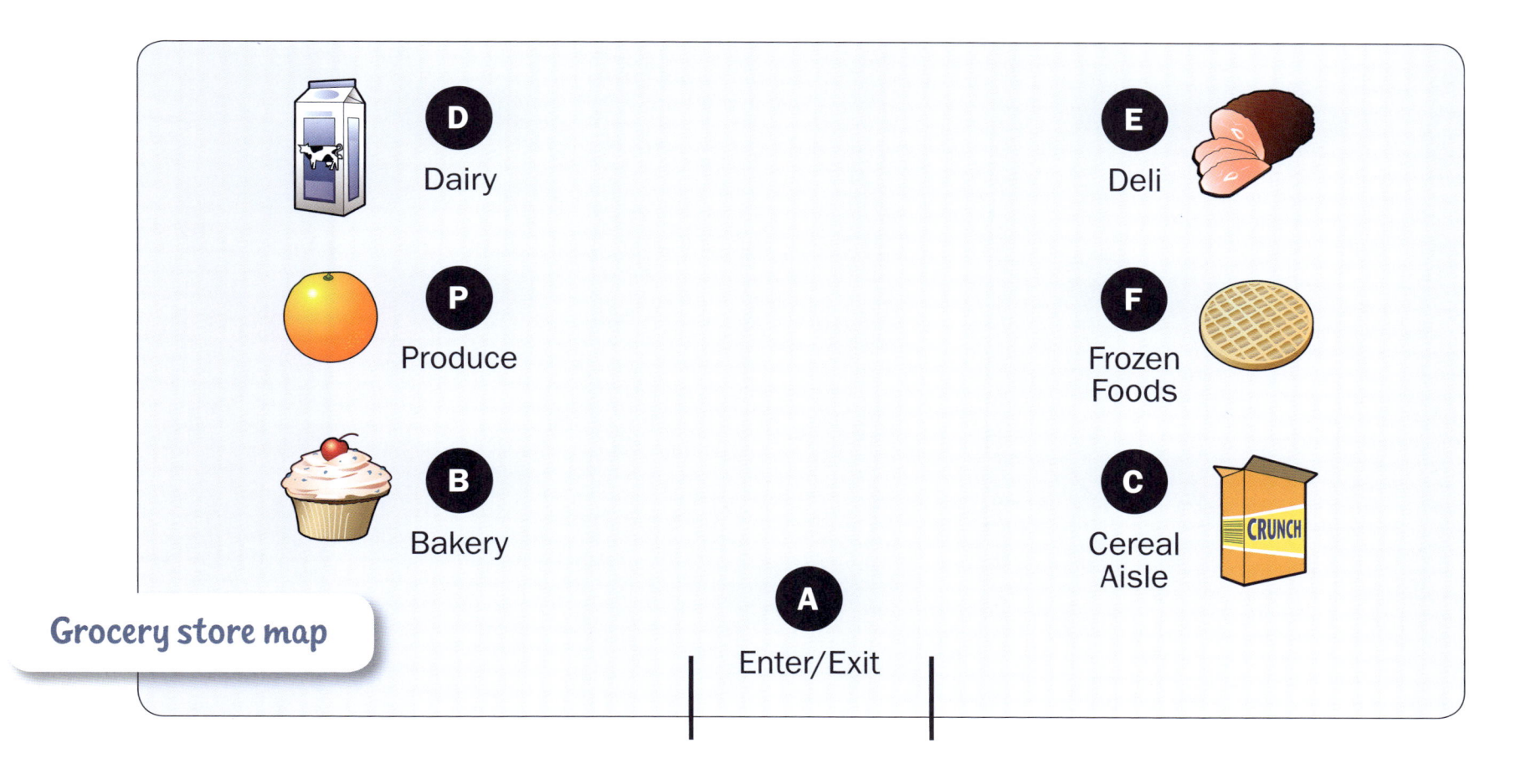

Grocery store map

- D — Dairy
- P — Produce
- B — Bakery
- E — Deli
- F — Frozen Foods
- C — Cereal Aisle
- A — Enter/Exit

Food: _____

Unit 1: Geometry

UNIT 2 ALGEBRA

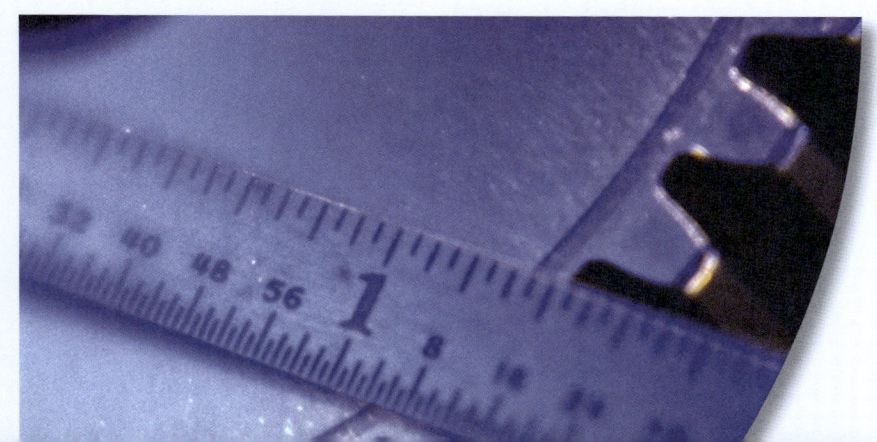

Time to shop

Ben arrived at South Park Mall at 2 o'clock.

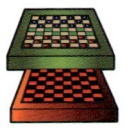

He shopped for clothes and a game at his favorite stores. His dad picked him up  at 5 o'clock, after Ben finished shopping.

How many hours did Ben shop?

Story 1: Time to shop

WHAT DO WE NEED TO FIND OUT? CHECK THE BOX. ☑

☐ 1. What game did Ben buy?

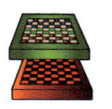

☐ 2. How many hours did Ben shop?

First fact	Sign	Second fact	Sign	Last fact
	+ −		=	

1 2 3 4 5 6 7 8 9 10

Add →   ← Subtract

Equation prompt

X = _____

Unit 2: Algebra

At the mall with Jean

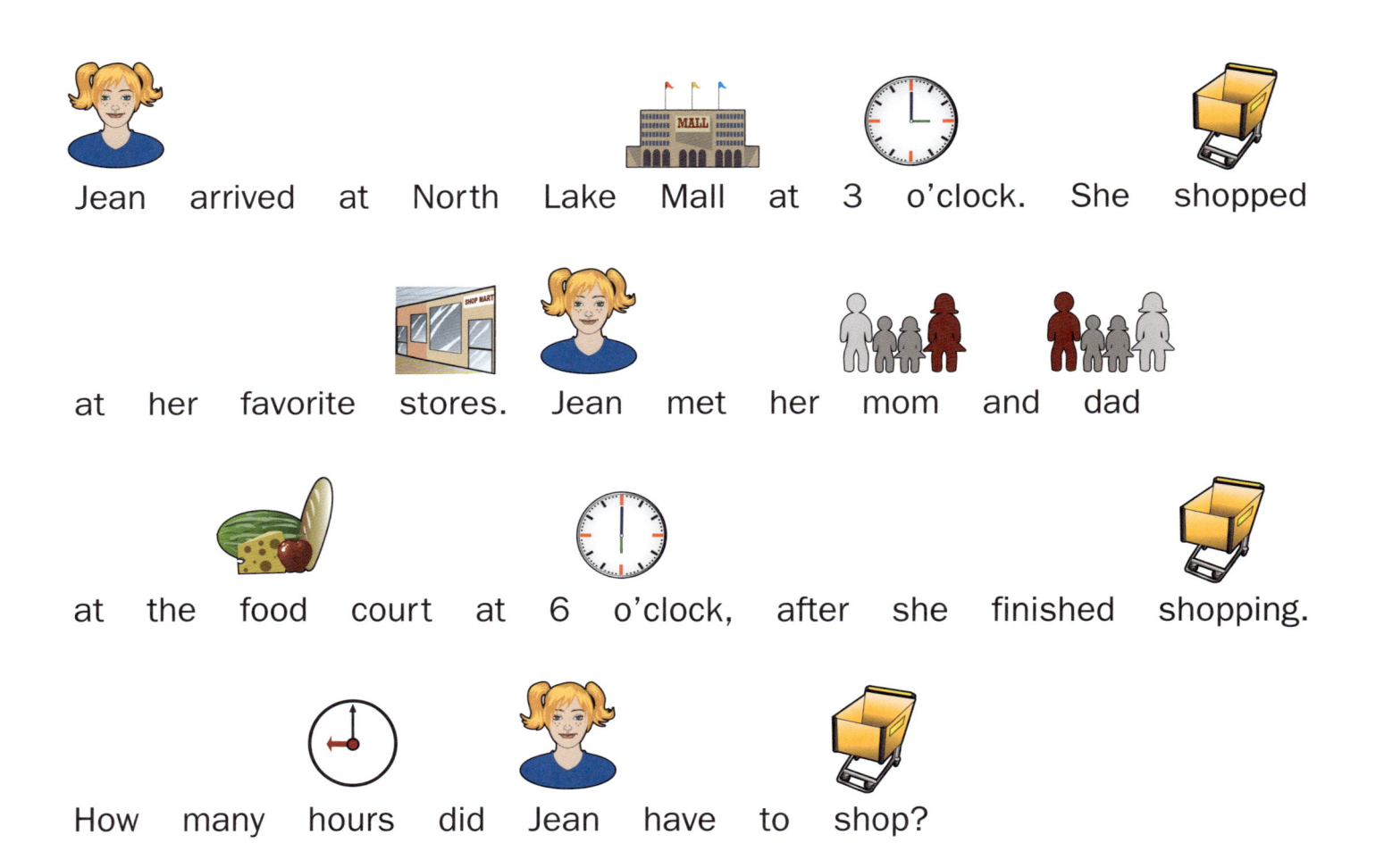

Jean arrived at North Lake Mall at 3 o'clock. She shopped at her favorite stores. Jean met her mom and dad at the food court at 6 o'clock, after she finished shopping.

How many hours did Jean have to shop?

46

Story 2: At the mall with Jean

WHAT DO WE NEED TO FIND OUT? CHECK THE BOX. ☑

☐ 1. How many hours did Jean have to shop?

☐ 2. What is Jean's favorite store?

First fact	Sign	Second fact	Sign	Last fact
	+ −		=	

1 2 3 4 5 6 7 8 9 10

Add ⟶   ⟵ Subtract

Equation prompt

X = _____

Unit 2: Algebra

47

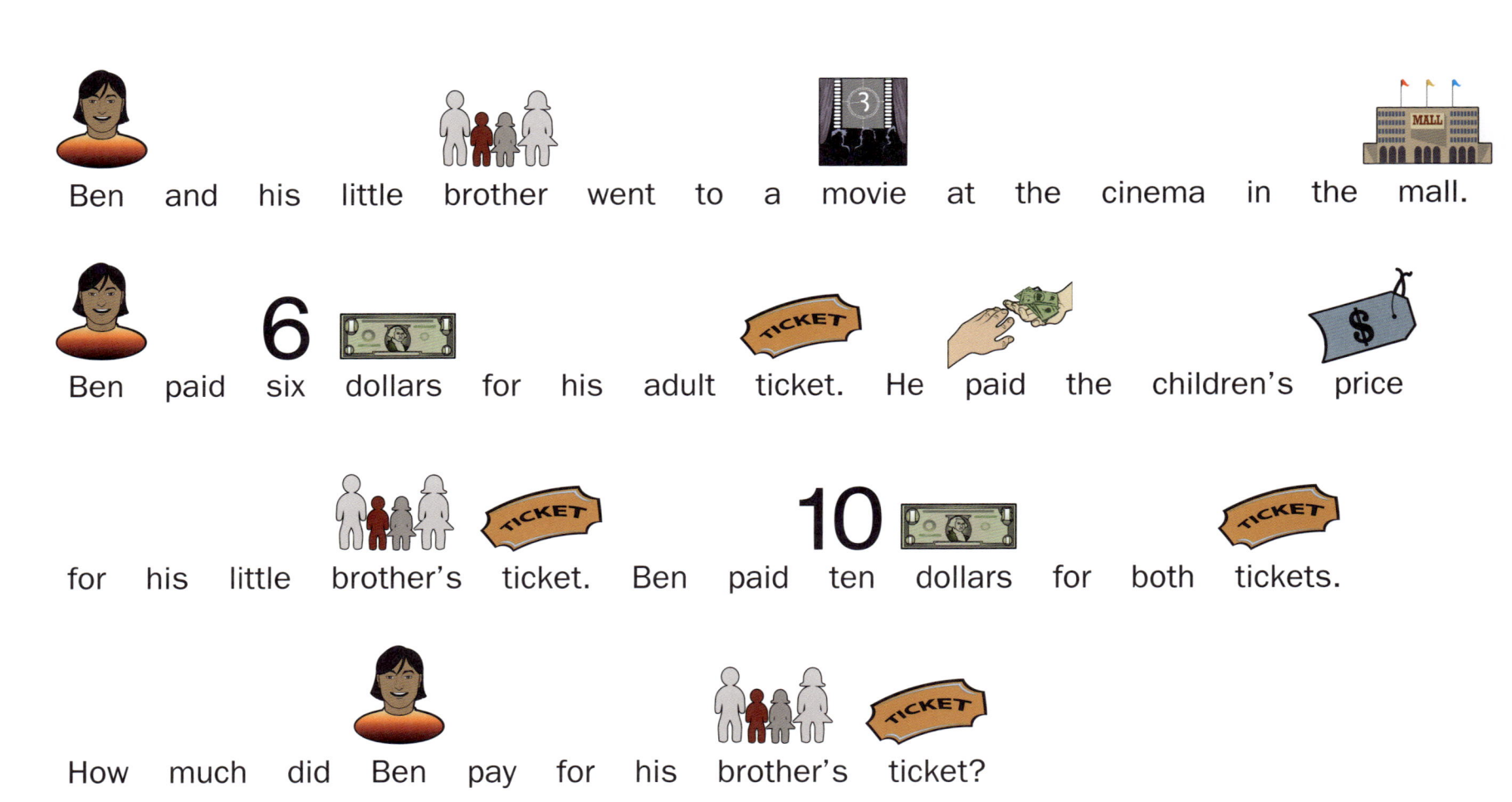

A movie for Ben and his brother

Ben and his little brother went to a movie at the cinema in the mall.

Ben paid six dollars for his adult ticket. He paid the children's price for his little brother's ticket. Ben paid ten dollars for both tickets.

How much did Ben pay for his brother's ticket?

WHAT DO WE NEED TO FIND OUT? CHECK THE BOX. ☑

☐ 1. What movie did Ben and his brother see?

☐ 2. How much did Ben pay for his brother's ticket?

First fact	Sign	Second fact	Sign	Last fact
	+ −		=	

1 2 3 4 5 6 7 8 9 10

Add ⟶   ⟵ Subtract

Equation prompt

X = _____

Unit 2: Algebra   49

Pavel works up an appetite

Pavel went shopping at the mall. He was hungry, so he went to the food court. He bought a sandwich for four dollars.

He also bought soup. Pavel spent eight dollars on his lunch.

How many dollars did Pavel spend on his soup?

WHAT DO WE NEED TO FIND OUT? CHECK THE BOX. ☑

☐ 1. How many dollars did Pavel spend on his soup?

☐ 2. What kind of sandwich did Pavel buy?

First fact	Sign	Second fact	Sign	Last fact
	+ −		=	

1   2   3   4   5   6   7   8   9   10

Add ⟶   ⟵ Subtract

Equation prompt

X = _____

Unit 2: Algebra

Gift cards for the movies

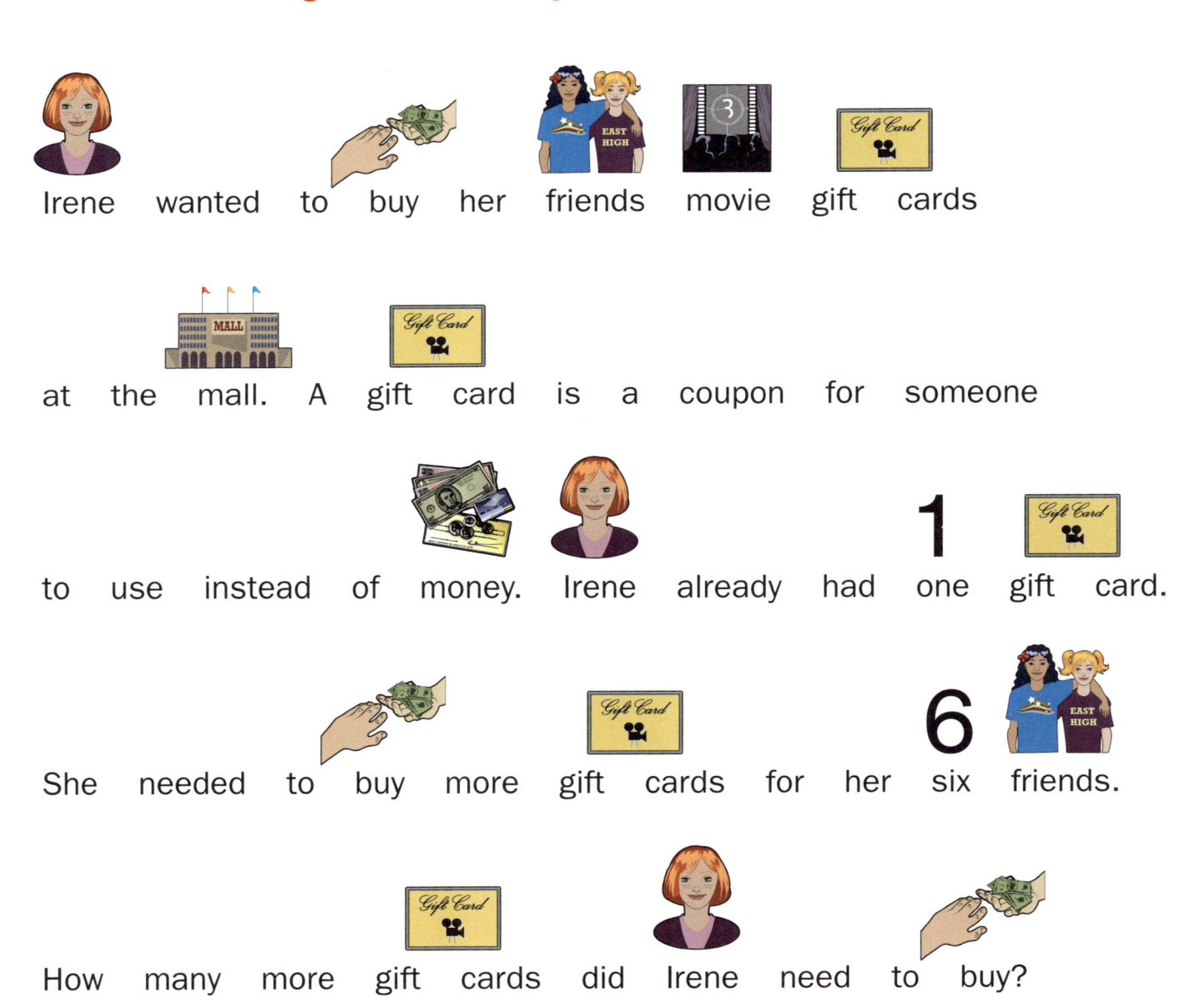

Irene wanted to buy her friends movie gift cards at the mall. A gift card is a coupon for someone to use instead of money. Irene already had one gift card. She needed to buy more gift cards for her six friends. How many more gift cards did Irene need to buy?

WHAT DO WE NEED TO FIND OUT? CHECK THE BOX. ☑

☐ 1. How many more gift cards did Irene need to buy?

☐ 2. Where does Irene shop for gifts?

First fact	Sign	Second fact	Sign	Last fact
	+ −		=	

1 2 3 4 5 6 7 8 9 10

Add ⟶   ⟵ Subtract

Equation prompt

X = _____

Unit 2: Algebra 53

Dominick shows a movie

Dominick invited some friends to watch a movie at his house. Five friends from school came to his house. He invited friends from the gym, too. Altogether, Dominick had nine friends come to his house to watch the movie. How many friends from the gym came to Dominick's house?

WHAT DO WE NEED TO FIND OUT? CHECK THE BOX. ☑

☐ 1. How many friends from the gym came to Dominick's house?

☐ 2. What movie did Dominick and his friends watch?

First fact	Sign	Second fact	Sign	Last fact
	+ −		=	

1 2 3 4 5 6 7 8 9 10

Add ⟶   ⟵ Subtract

Equation prompt

X = _____

Unit 2: Algebra

55

Maria's house party

Maria invited some friends to watch a movie at her house. Three friends from school came to her house. She invited friends from the gym, too.

Altogether, Maria had eight friends come to watch the movie.

How many friends from the gym came to Maria's house?

WHAT DO WE NEED TO FIND OUT? CHECK THE BOX. ✓

☐ 1. How many pizzas did Maria and her friends eat?

☐ 2. How many friends from the gym came to Maria's house?

First fact	Sign	Second fact	Sign	Last fact
	+ −		=	

1   2   3   4   5   6   7   8   9   10

Add ⟶   ⟵ Subtract

Equation prompt

X = _____

Unit 2: Algebra

57

Movie time

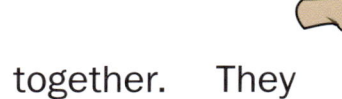

Maria and her friends watched a movie together. They started the movie at 3 o'clock. They watched the whole movie without stopping it. The movie ended at 5 o'clock. How long was the movie?

WHAT DO WE NEED TO FIND OUT? CHECK THE BOX. ✓

☐ 1. How long was the movie?

☐ 2. How many friends watched the movie with Maria?

First fact	Sign	Second fact	Sign	Last fact
	+ −		=	

1 2 3 4 5 6 7 8 9 10

Add ⟶   ⟵ Subtract

Equation prompt

X = _____

Unit 2: Algebra

59

Let's order pizza

Randy and his friends are having a party. They ordered cheese and pepperoni pizza for dinner. They ordered two cheese pizzas. They ordered some pepperoni pizza, too. Altogether, they ordered nine pizzas. How many pepperoni pizzas did Randy and his friends order?

WHAT DO WE NEED TO FIND OUT? CHECK THE BOX. ☑

☐ 1. How much did Randy and his friends spend on pizzas?

☐ 2. How many pepperoni pizzas did Randy and his friends order?

First fact	Sign	Second fact	Sign	Last fact
	+ −		=	

1 2 3 4 5 6 7 8 9 10

Add →   ← Subtract

Equation prompt

X = _____

Unit 2: Algebra

61

Paying for dinner

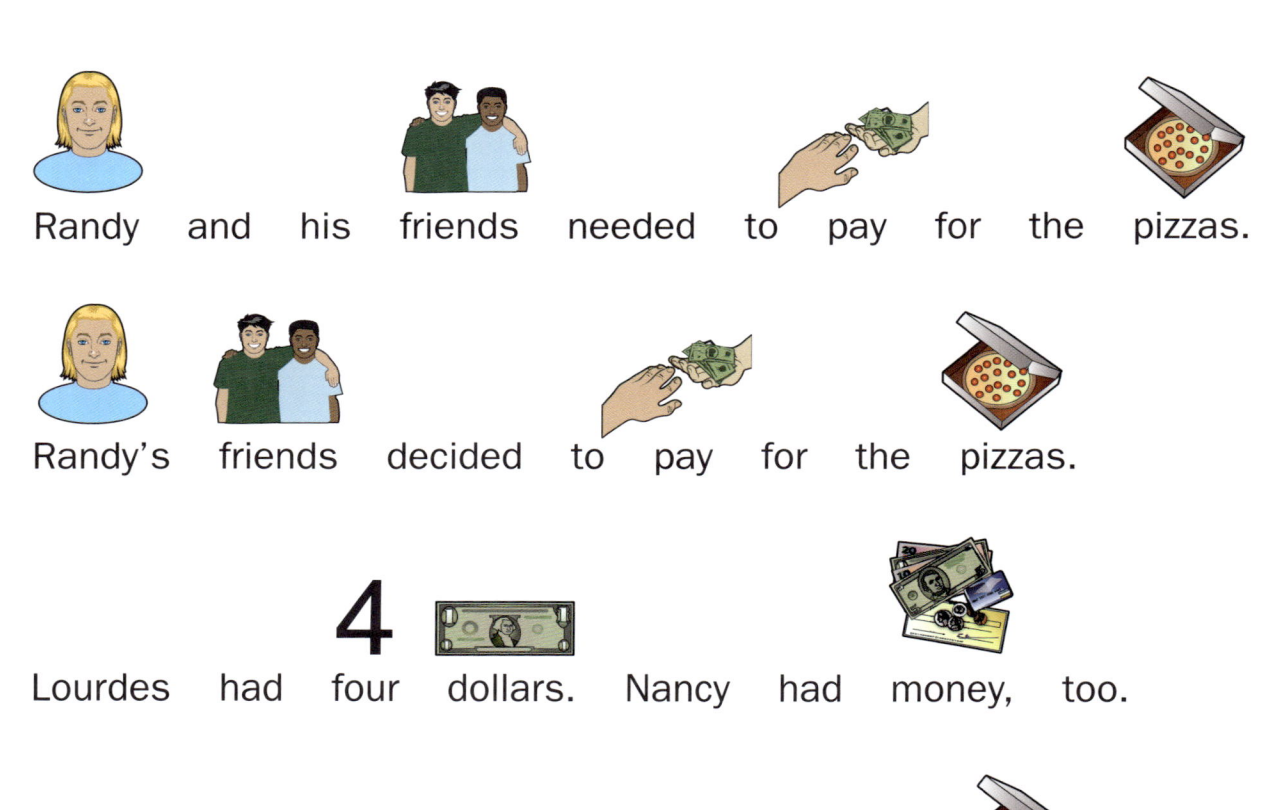

Randy and his friends needed to pay for the pizzas.

Randy's friends decided to pay for the pizzas.

Lourdes had four dollars. Nancy had money, too.

Altogether, they paid ten dollars for the pizzas.

How many dollars did Nancy pay for the pizzas?

WHAT DO WE NEED TO FIND OUT? CHECK THE BOX. ☑

☐ 1. What kind of pizza do Randy's friends like?

☐ 2. How many dollars did Nancy pay for the pizzas?

First fact	Sign	Second fact	Sign	Last fact
	+ −		=	

1   2   3   4   5   6   7   8   9   10

Add ⟶   ⟵ Subtract

Equation prompt

X = _____

Unit 2: Algebra

63

Ms. Black's class choices

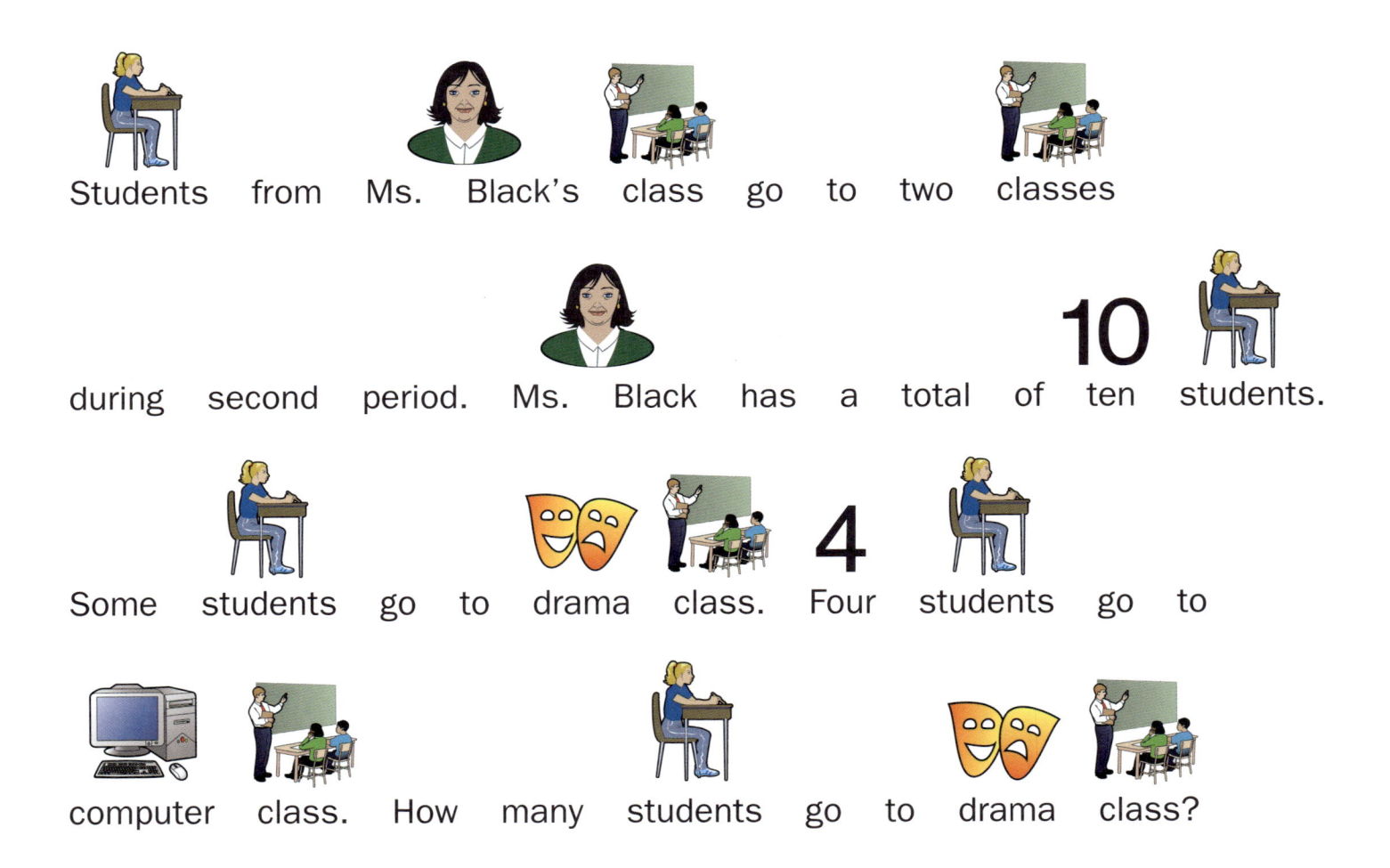

Students from Ms. Black's class go to two classes during second period. Ms. Black has a total of ten students. Some students go to drama class. Four students go to computer class. How many students go to drama class?

WHAT DO WE NEED TO FIND OUT? CHECK THE BOX. ☑

☐ 1. How many students go to drama class?

☐ 2. What does Ms. Black teach?

First fact	Sign	Second fact	Sign	Last fact
	+ −		=	

1 2 3 4 5 6 7 8 9 10

Add ⟶   ⟵ Subtract

Equation prompt

X = _____

Unit 2: Algebra

65

Mr. Williams's students go to class

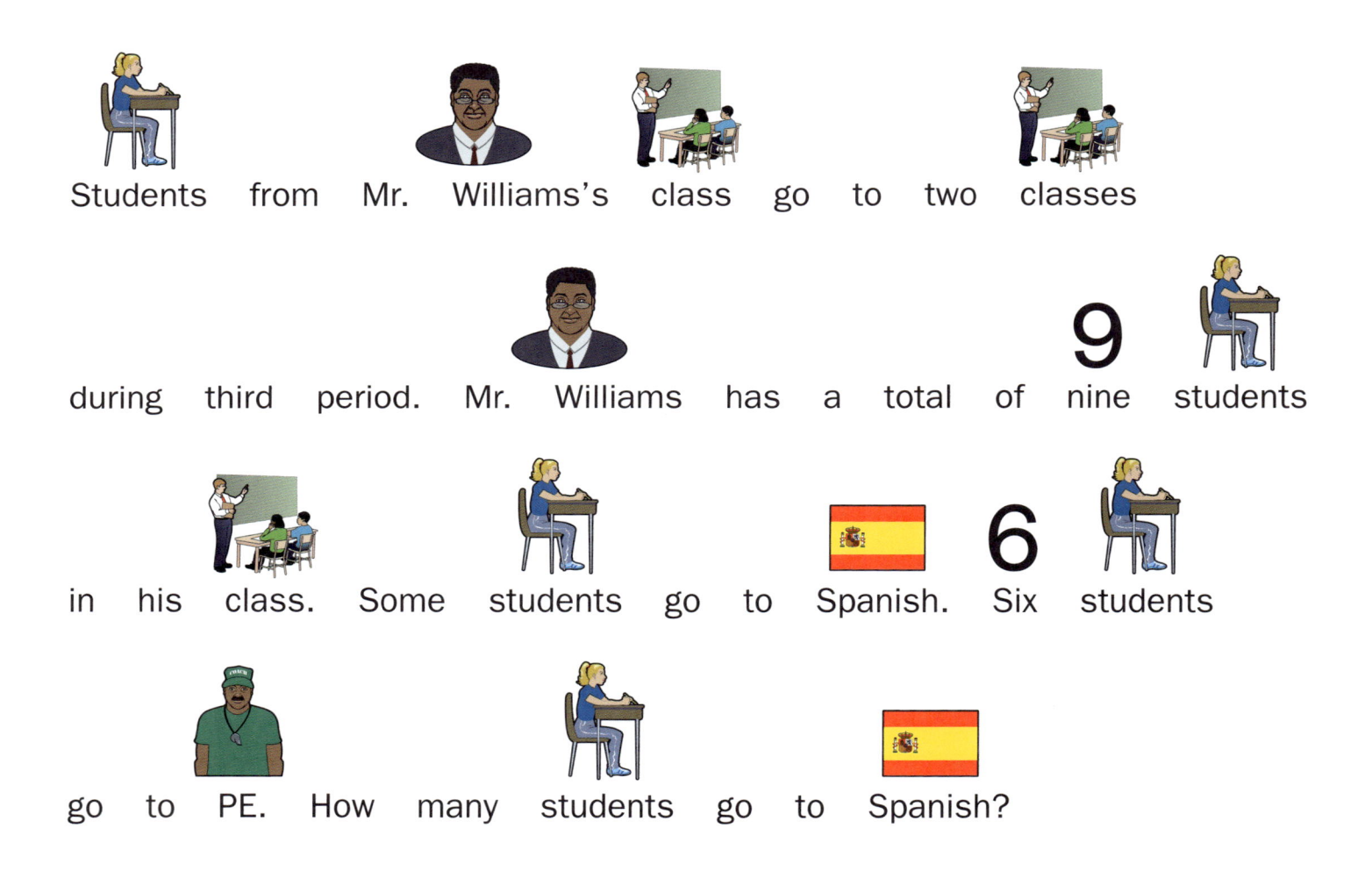

Students from Mr. Williams's class go to two classes during third period. Mr. Williams has a total of 9 nine students in his class. Some students go to Spanish. Six students go to PE. How many students go to Spanish?

WHAT DO WE NEED TO FIND OUT? CHECK THE BOX. ✓

☐ 1. How old are the students in Mr. Williams's class?

☐ 2. How many students go to Spanish?

First fact	Sign	Second fact	Sign	Last fact
	+ –		=	

1 2 3 4 5 6 7 8 9 10

Add ⟶   ⟵ Subtract

Equation prompt

X = _____

Unit 2: Algebra

67

Mr. Fig's class chooses pizza

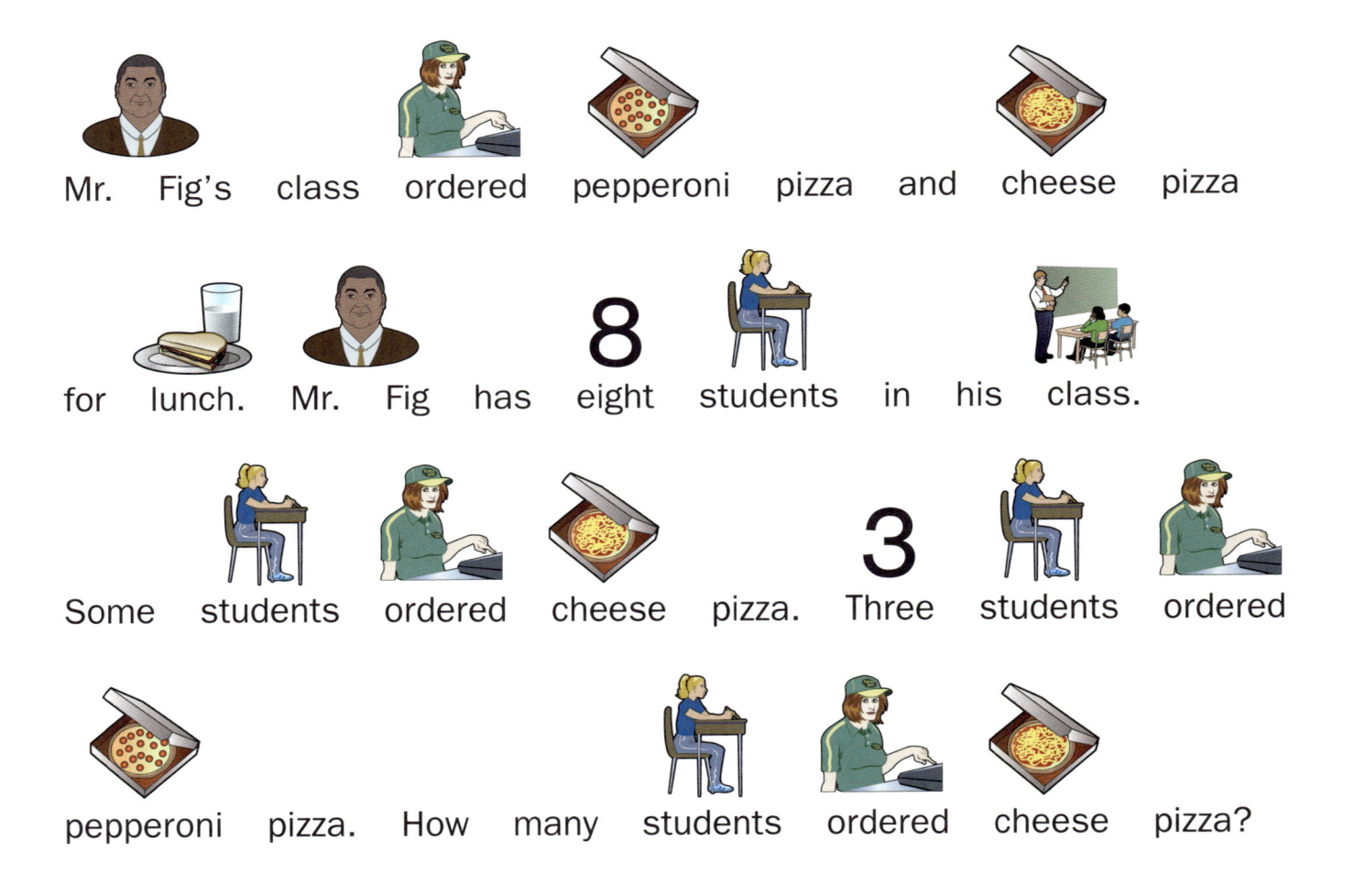

Mr. Fig's class ordered pepperoni pizza and cheese pizza for lunch. Mr. Fig has 8 eight students in his class. Some students ordered cheese pizza. Three 3 students ordered pepperoni pizza. How many students ordered cheese pizza?

WHAT DO WE NEED TO FIND OUT? CHECK THE BOX. ☑

☐ 1. How many students ordered juice?

☐ 2. How many students ordered cheese pizza?

First fact	Sign	Second fact	Sign	Last fact
	+ −		=	

1 2 3 4 5 6 7 8 9 10

Add ⟶   ⟵ Subtract

Equation prompt

X = _____

Unit 2: Algebra

69

Computer class plans future

The students in computer class talk about office jobs.

There are a total of ten students in the class.

Some students want to work in a store office.

Eight students want to work in a bank office.

How many students want to work in a store office?

WHAT DO WE NEED TO FIND OUT? CHECK THE BOX. ☑

☐ 1. Who teaches the computer class?

☐ 2. How many students want to work in a store office?

First fact	Sign	Second fact	Sign	Last fact
	+ −		=	

• • • • • • • • • •
1 2 3 4 5 6 7 8 9 10

Add ⟶   ⟵ Subtract

Equation prompt

X = _____

Unit 2: Algebra

71

PE class plans summer fun

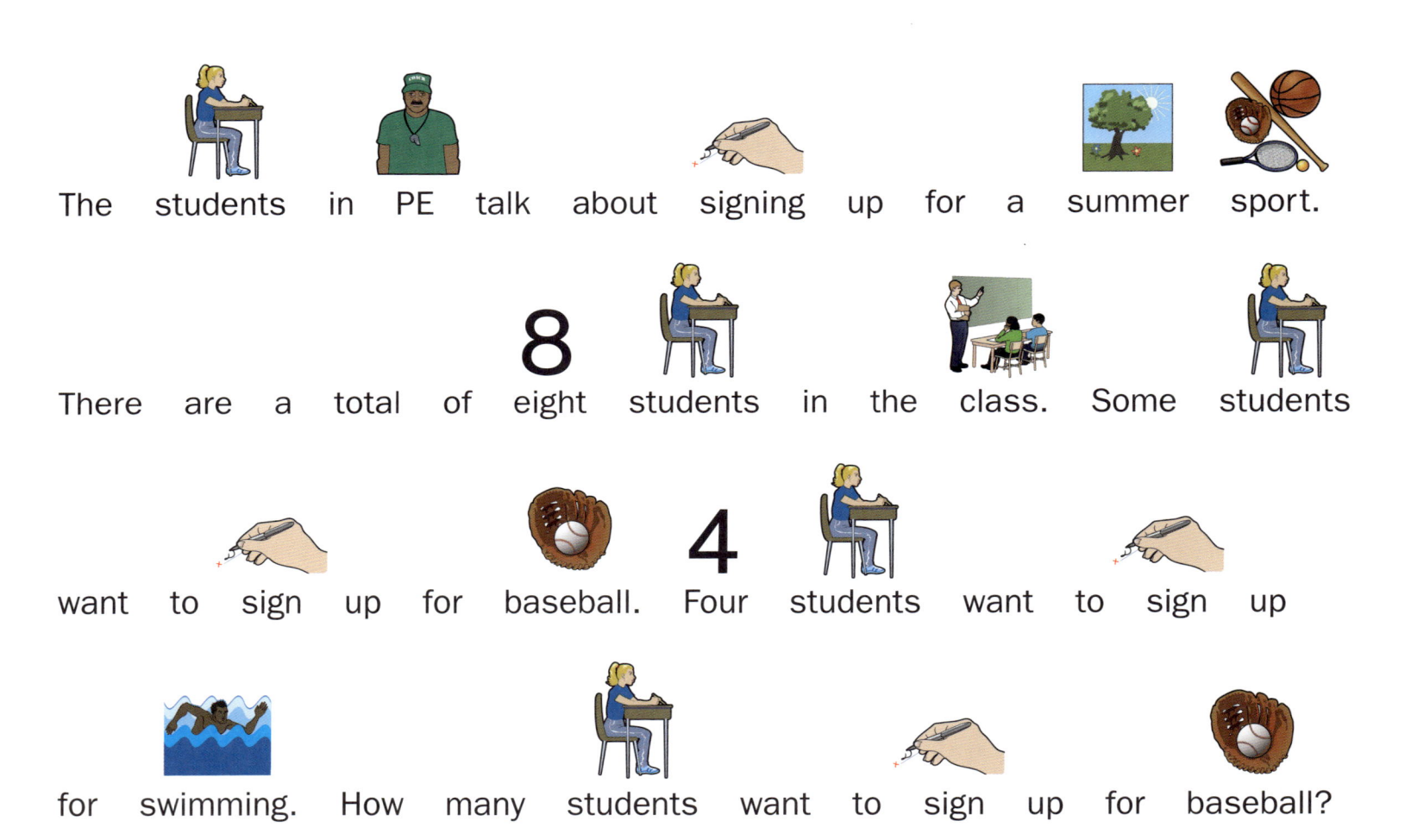

The students in PE talk about signing up for a summer sport. There are a total of eight students in the class. Some students want to sign up for baseball. Four students want to sign up for swimming. How many students want to sign up for baseball?

WHAT DO WE NEED TO FIND OUT? CHECK THE BOX. ☑

☐ 1. How many students want to sign up for baseball?

☐ 2. Who is the best swimmer in the class?

First fact	Sign	Second fact	Sign	Last fact
	+ −		=	

1   2   3   4   5   6   7   8   9   10

Add ⟶   ⟵ Subtract

Equation prompt

X = _____

Unit 2: Algebra

73

Spanish class makes travel plans

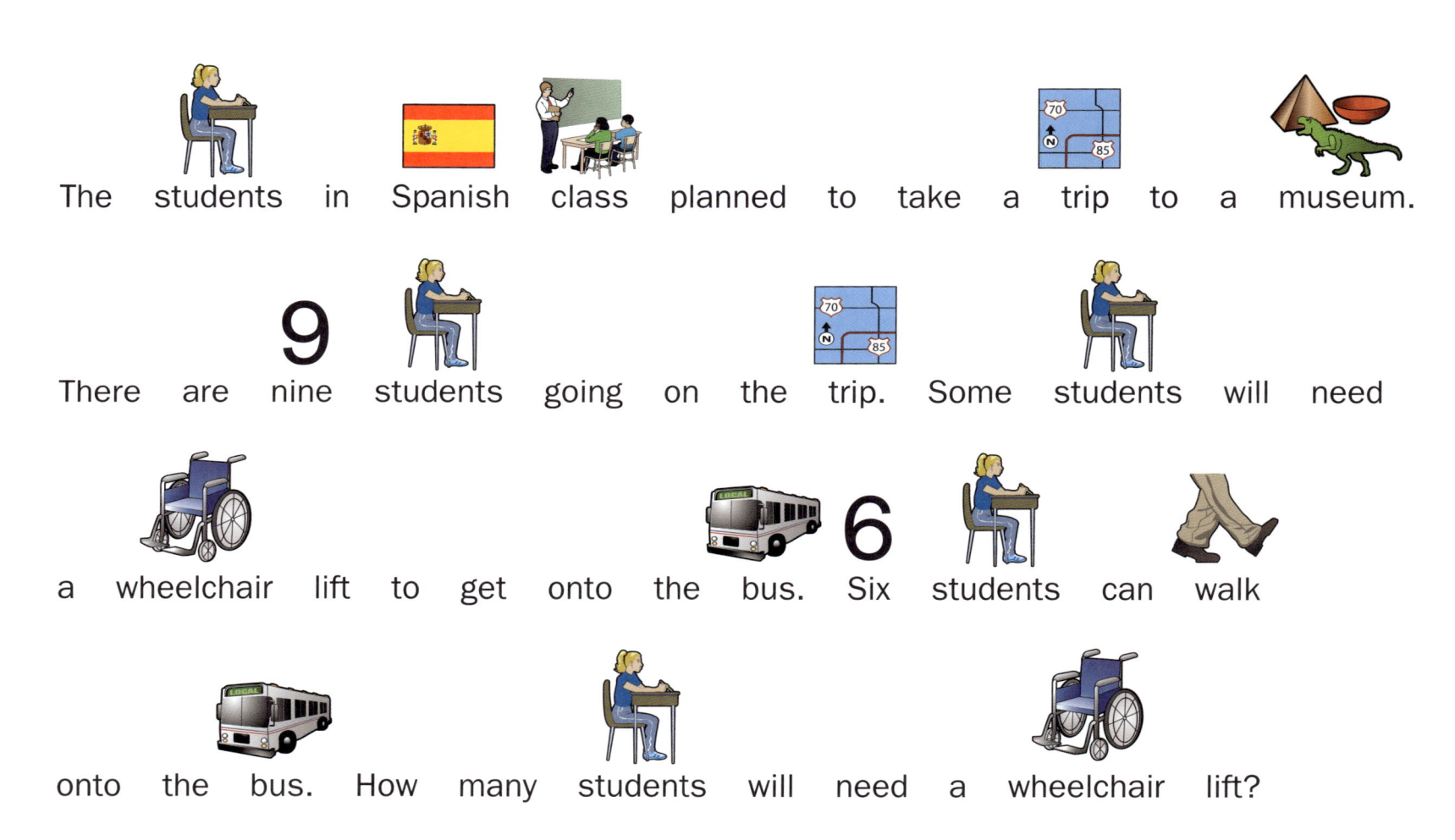

The students in Spanish class planned to take a trip to a museum. There are nine students going on the trip. Some students will need a wheelchair lift to get onto the bus. Six students can walk onto the bus. How many students will need a wheelchair lift?

WHAT DO WE NEED TO FIND OUT? CHECK THE BOX. ✓

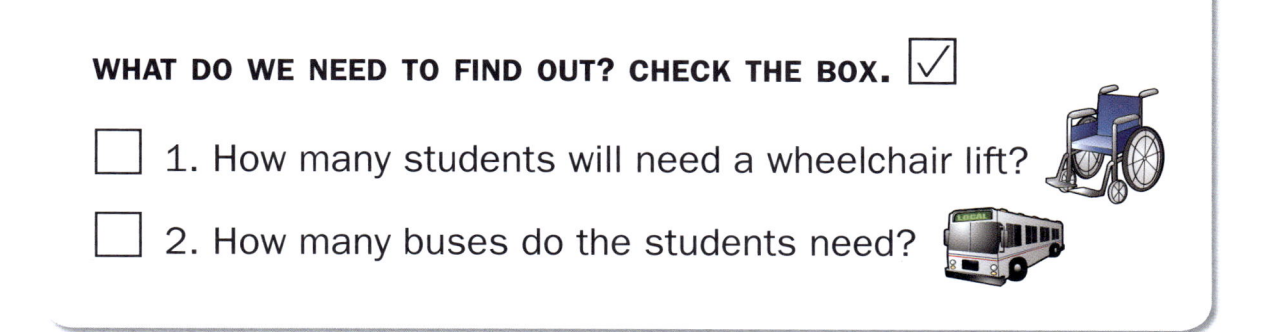

☐ 1. How many students will need a wheelchair lift?

☐ 2. How many buses do the students need?

First fact	Sign	Second fact	Sign	Last fact
	+ −		=	

1 2 3 4 5 6 7 8 9 10

Add ⟶   ⟵ Subtract

Equation prompt

X = _____

Unit 2: Algebra

75

Meagan at the food court

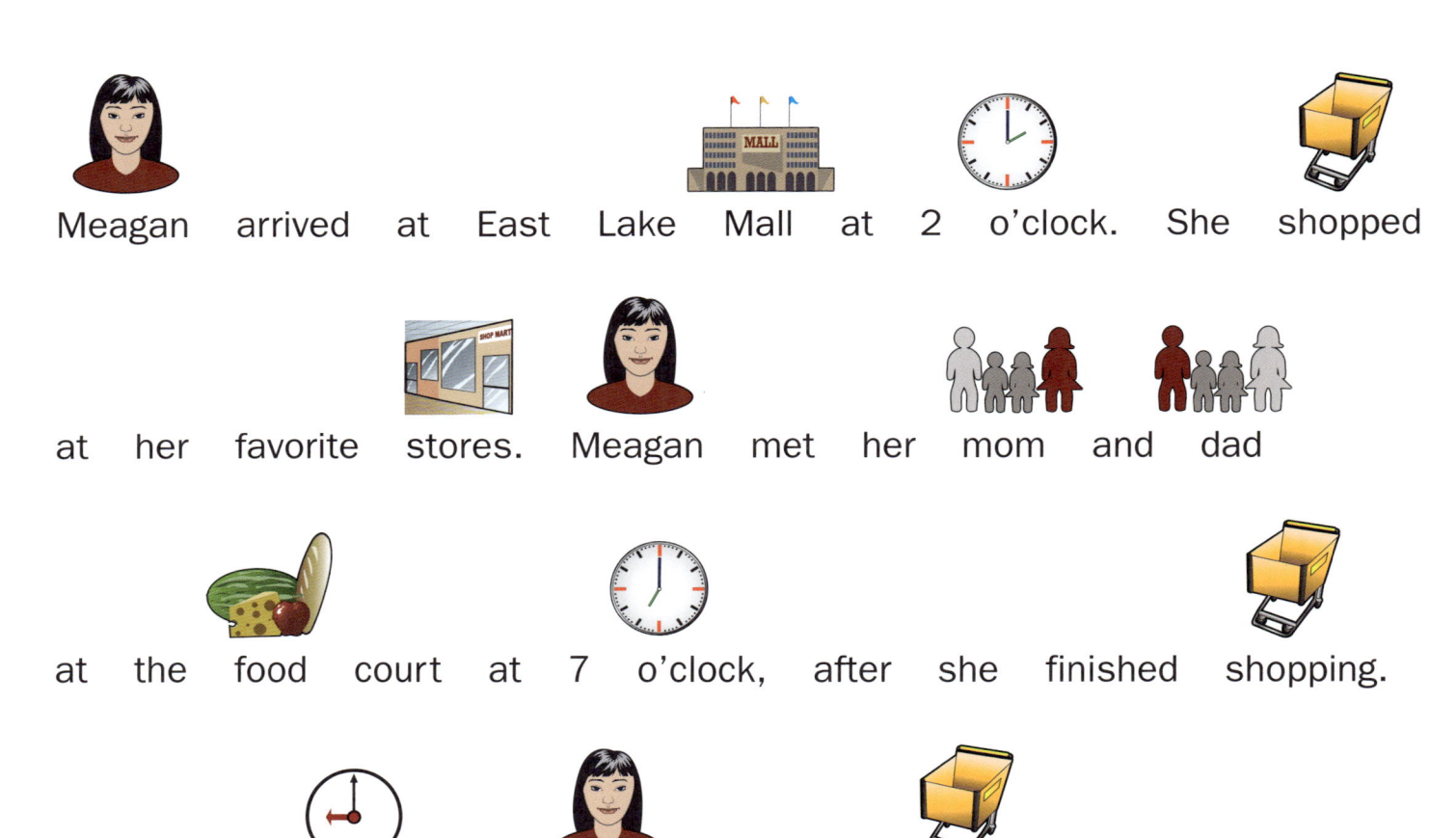

Meagan arrived at East Lake Mall at 2 o'clock. She shopped at her favorite stores. Meagan met her mom and dad at the food court at 7 o'clock, after she finished shopping.

How many hours did Meagan have to shop?

WHAT DO WE NEED TO FIND OUT? CHECK THE BOX. ✓

☐ 1. Where did Meagan go before she ate dinner?

☐ 2. How many hours did Meagan have to shop?

First fact	Sign	Second fact	Sign	Last fact
	+ −		=	

1 2 3 4 5 6 7 8 9 10

Add ⟶   ⟵ Subtract

Equation prompt

X = _____

Unit 2: Algebra

77

UNIT 3 DATA ANALYSIS

The Greens vote on a movie

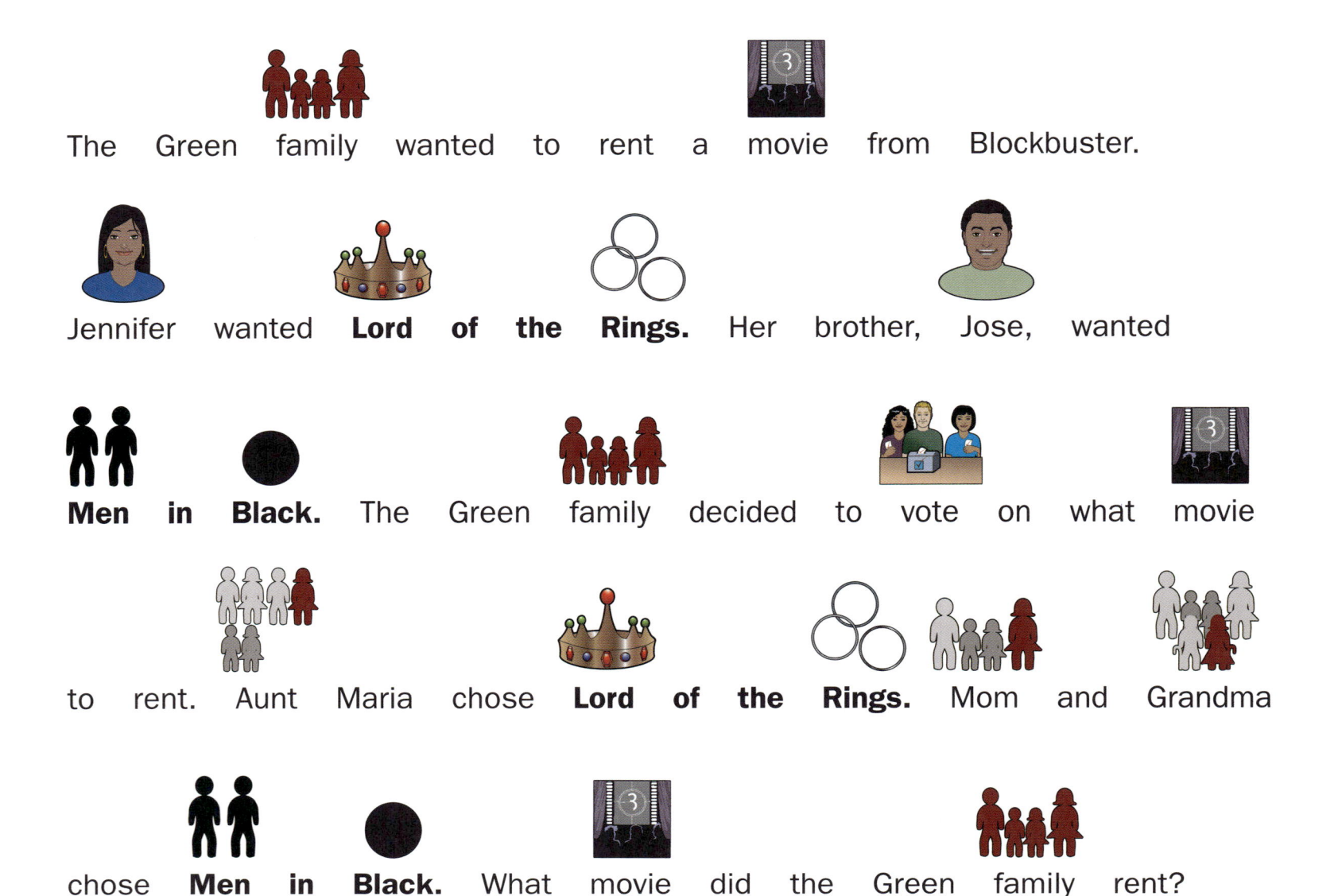

The Green family wanted to rent a movie from Blockbuster. Jennifer wanted **Lord of the Rings.** Her brother, Jose, wanted **Men in Black.** The Green family decided to vote on what movie to rent. Aunt Maria chose **Lord of the Rings.** Mom and Grandma chose **Men in Black.** What movie did the Green family rent?

WHAT DO WE NEED TO FIND OUT? CHECK THE BOX. ✓

☐ 1. What movie did the Green family rent?

☐ 2. Where did the Green family shop?

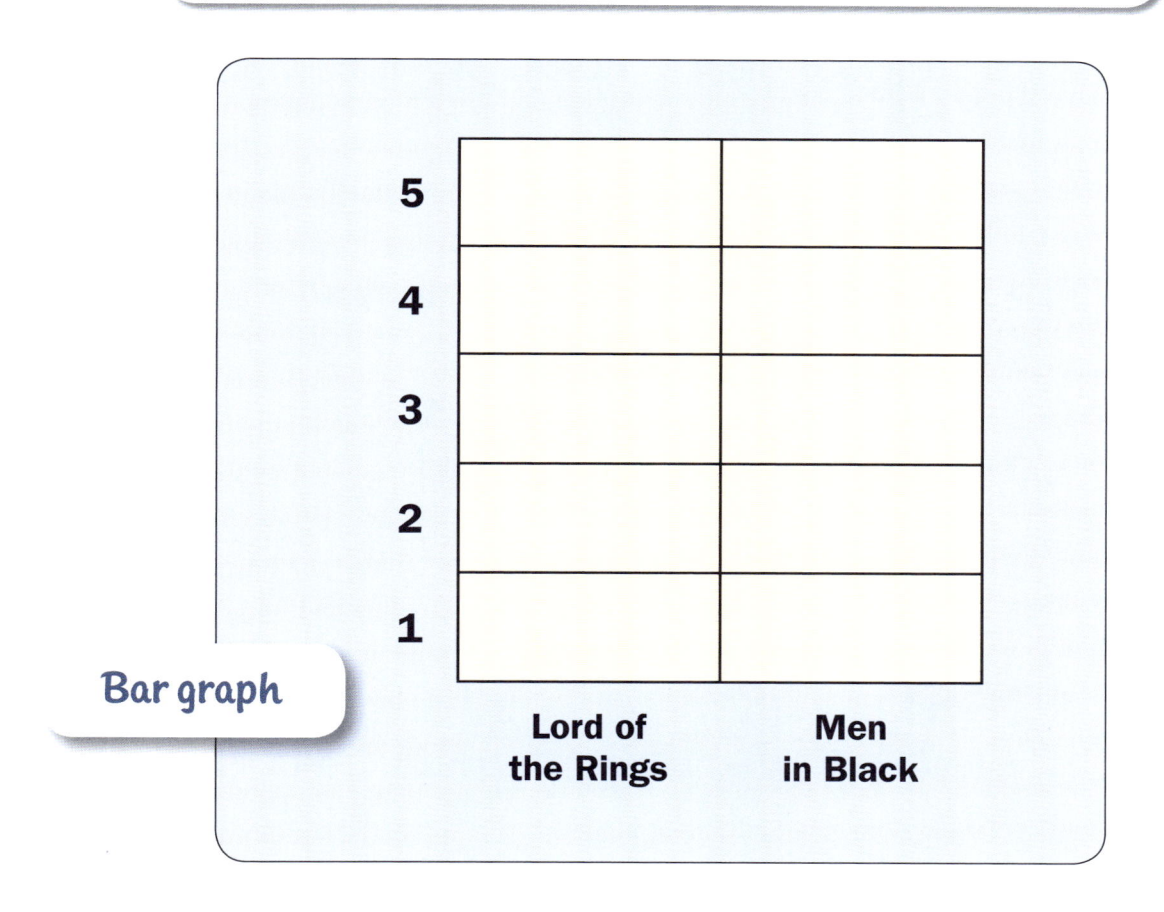

Bar graph

	Lord of the Rings	Men in Black
5		
4		
3		
2		
1		

Movie: _____

Mr. Taylor's class chooses a movie

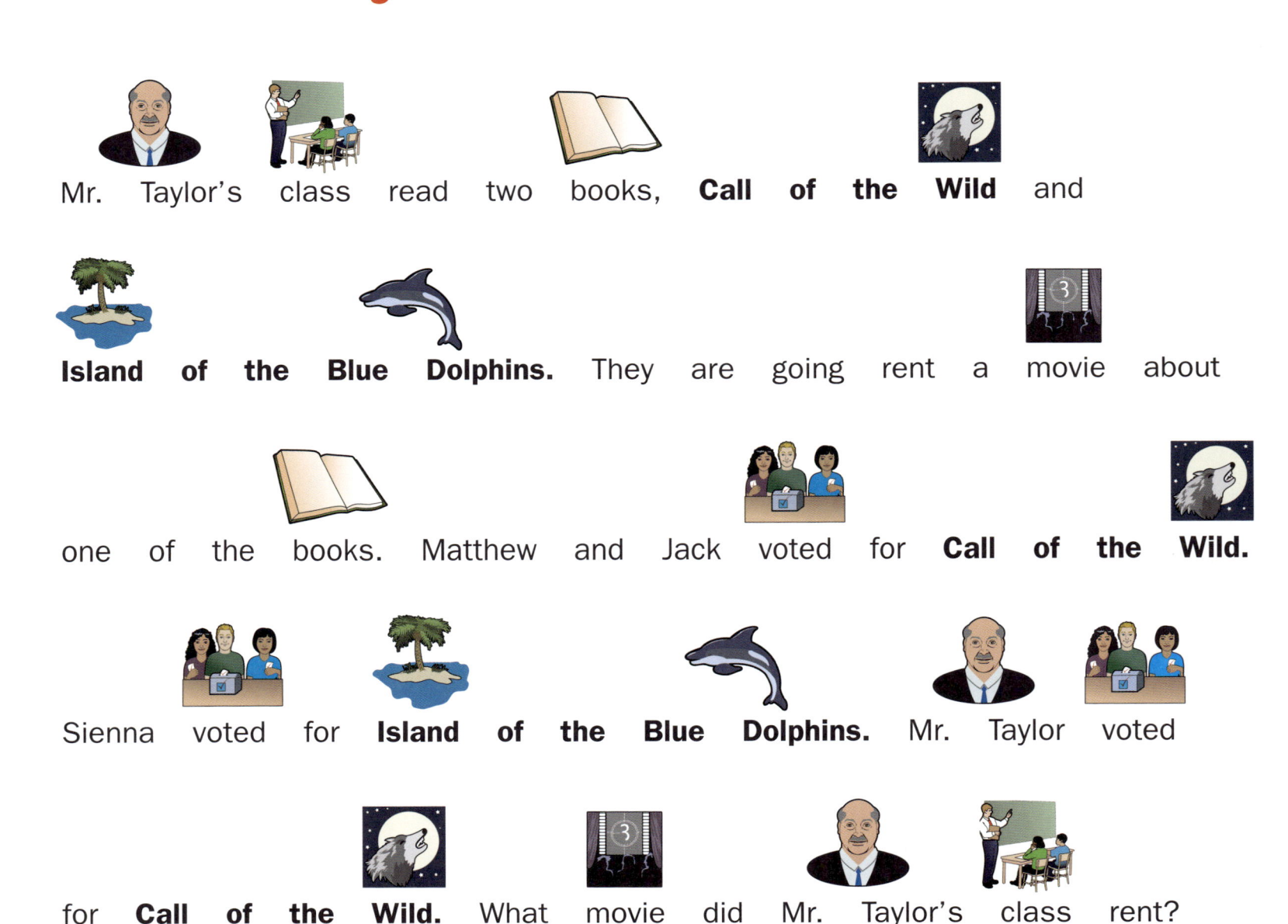

Mr. Taylor's class read two books, **Call of the Wild** and **Island of the Blue Dolphins.** They are going rent a movie about one of the books. Matthew and Jack voted for **Call of the Wild.** Sienna voted for **Island of the Blue Dolphins.** Mr. Taylor voted for **Call of the Wild.** What movie did Mr. Taylor's class rent?

WHAT DO WE NEED TO FIND OUT? CHECK THE BOX. ☑

☐ 1. How many books has Mr. Taylor's class read this year?

☐ 2. What movie did Mr. Taylor's class rent?

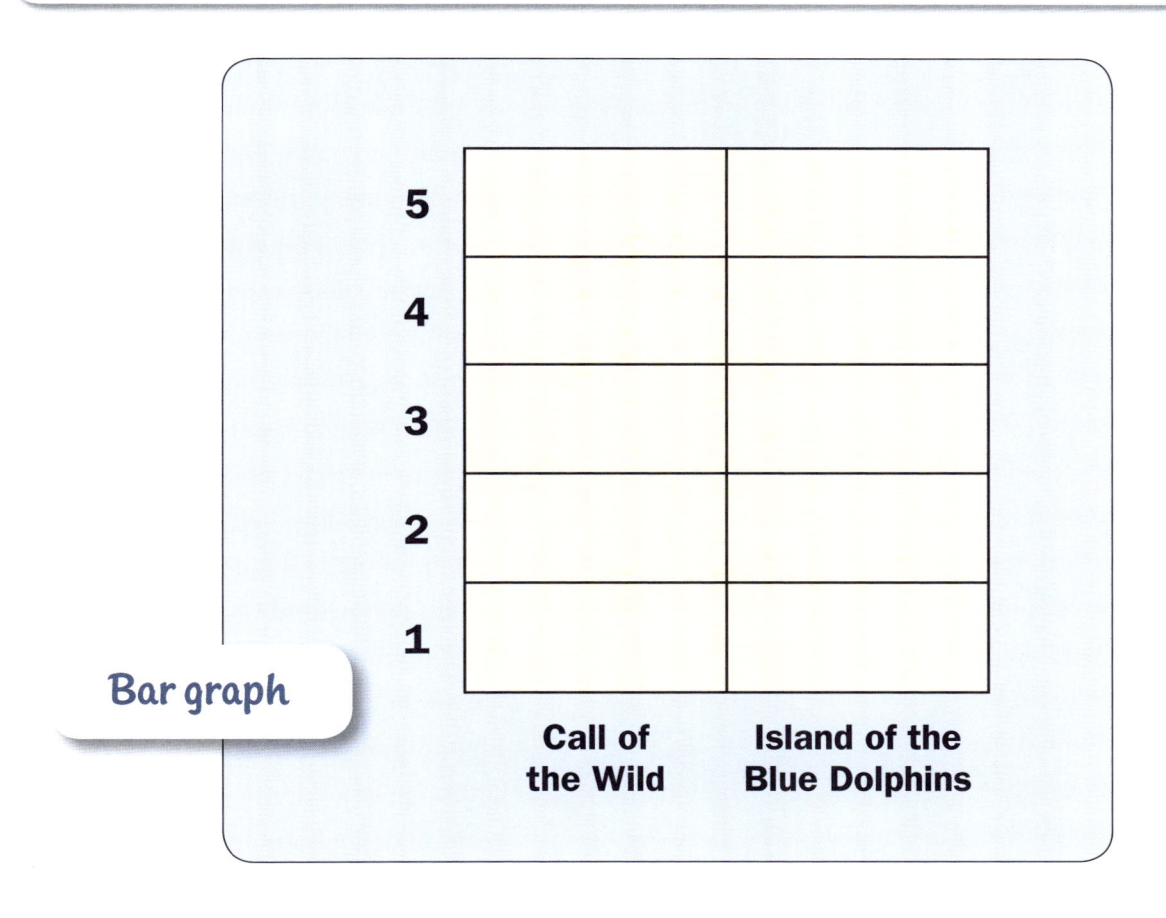

Bar graph

	Call of the Wild	Island of the Blue Dolphins
5		
4		
3		
2		
1		

Movie: _____

Unit 3: Data Analysis

83

Irene's friends choose a music video

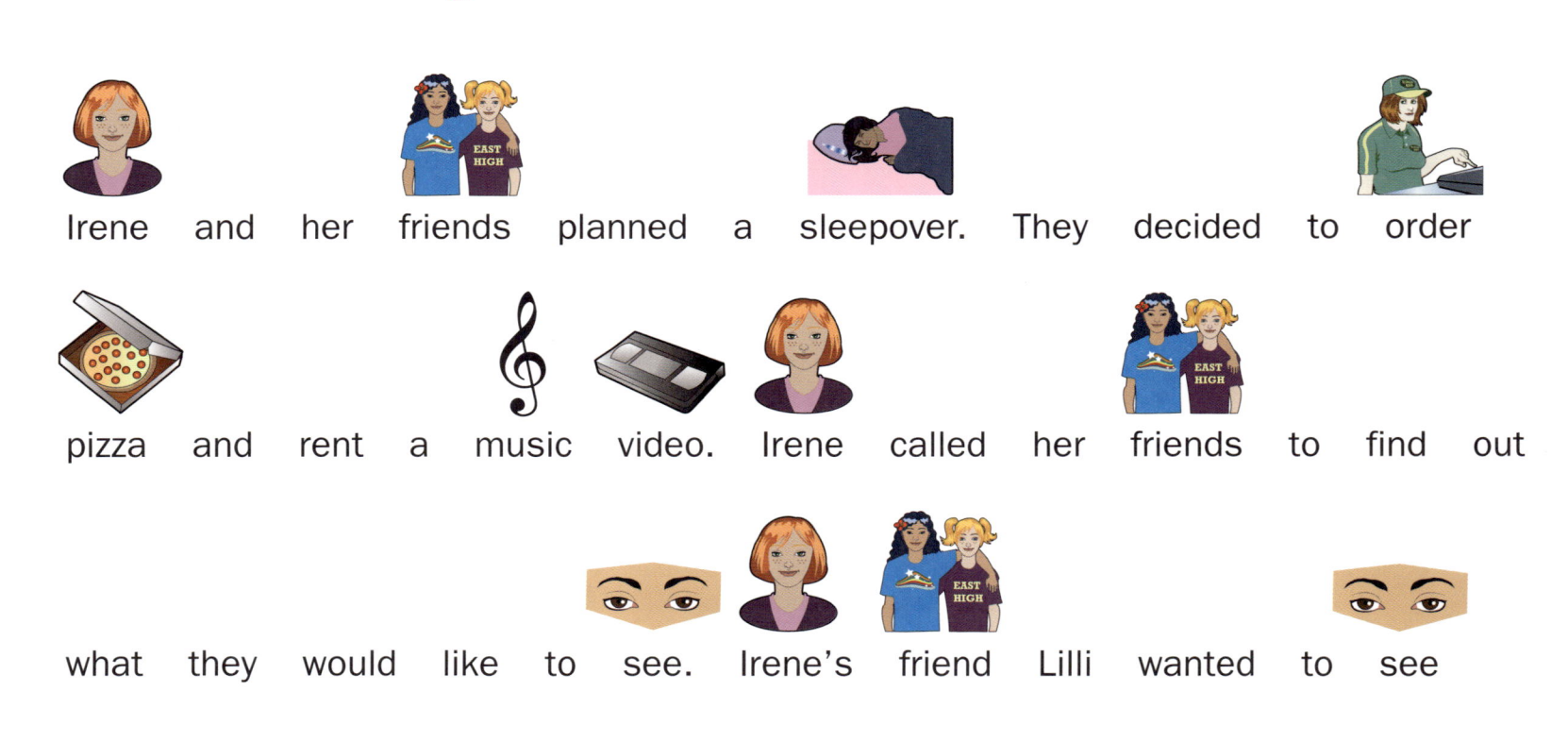

Irene and her friends planned a sleepover. They decided to order pizza and rent a music video. Irene called her friends to find out what they would like to see. Irene's friend Lilli wanted to see **Justin Timberlake.** Rose, Nicole, and Irene wanted to see **Hilary Duff.** Sally voted for **Bow Wow.** What music video did the friends rent?

WHAT DO WE NEED TO FIND OUT? CHECK THE BOX. ☑

☐ 1. What music video did the friends rent?

☐ 2. What is Irene's favorite snack?

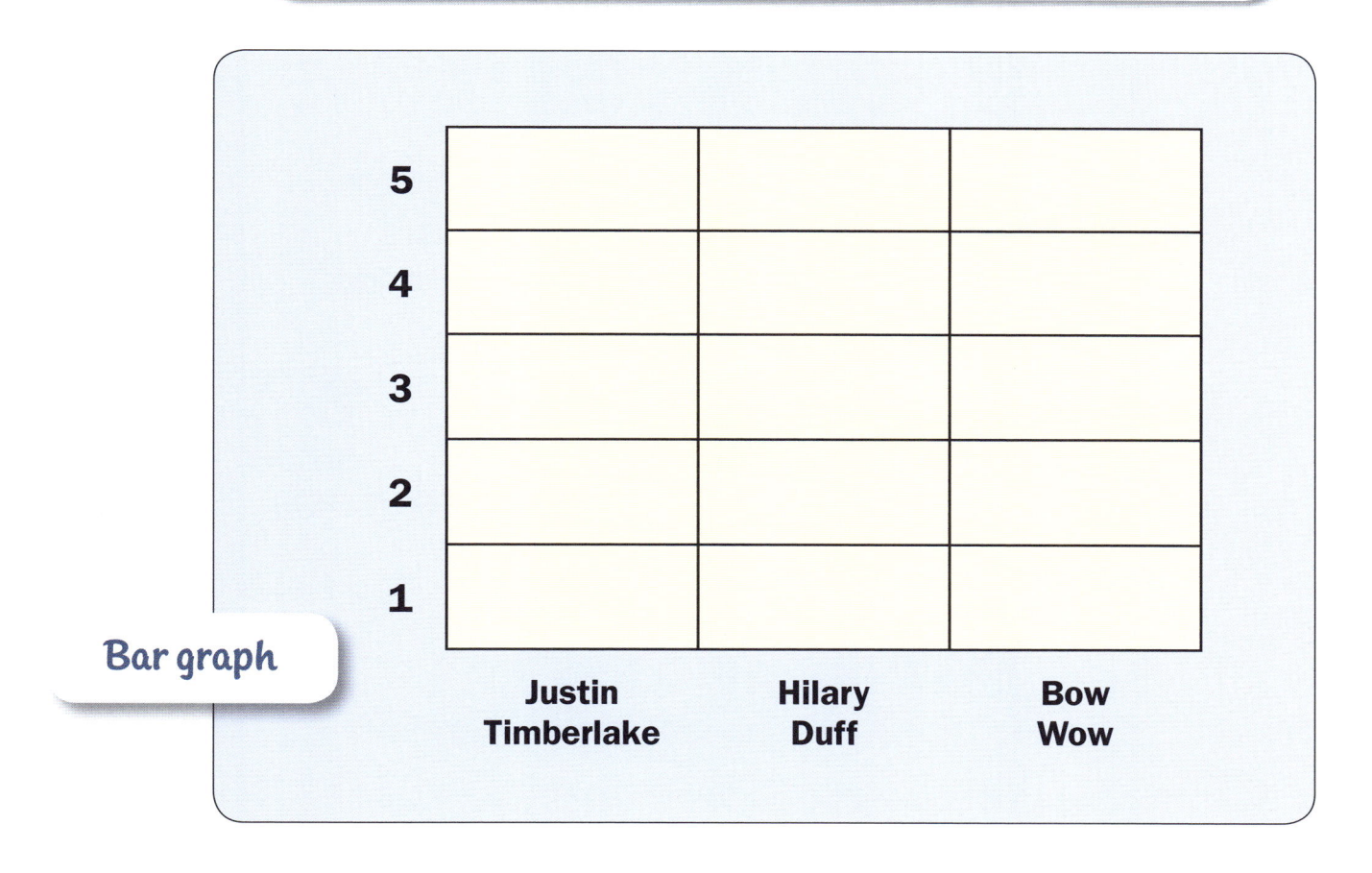

Bar graph

5			
4			
3			
2			
1			
	Justin Timberlake	Hilary Duff	Bow Wow

Video: _____

Unit 3: Data Analysis

Which video to watch?

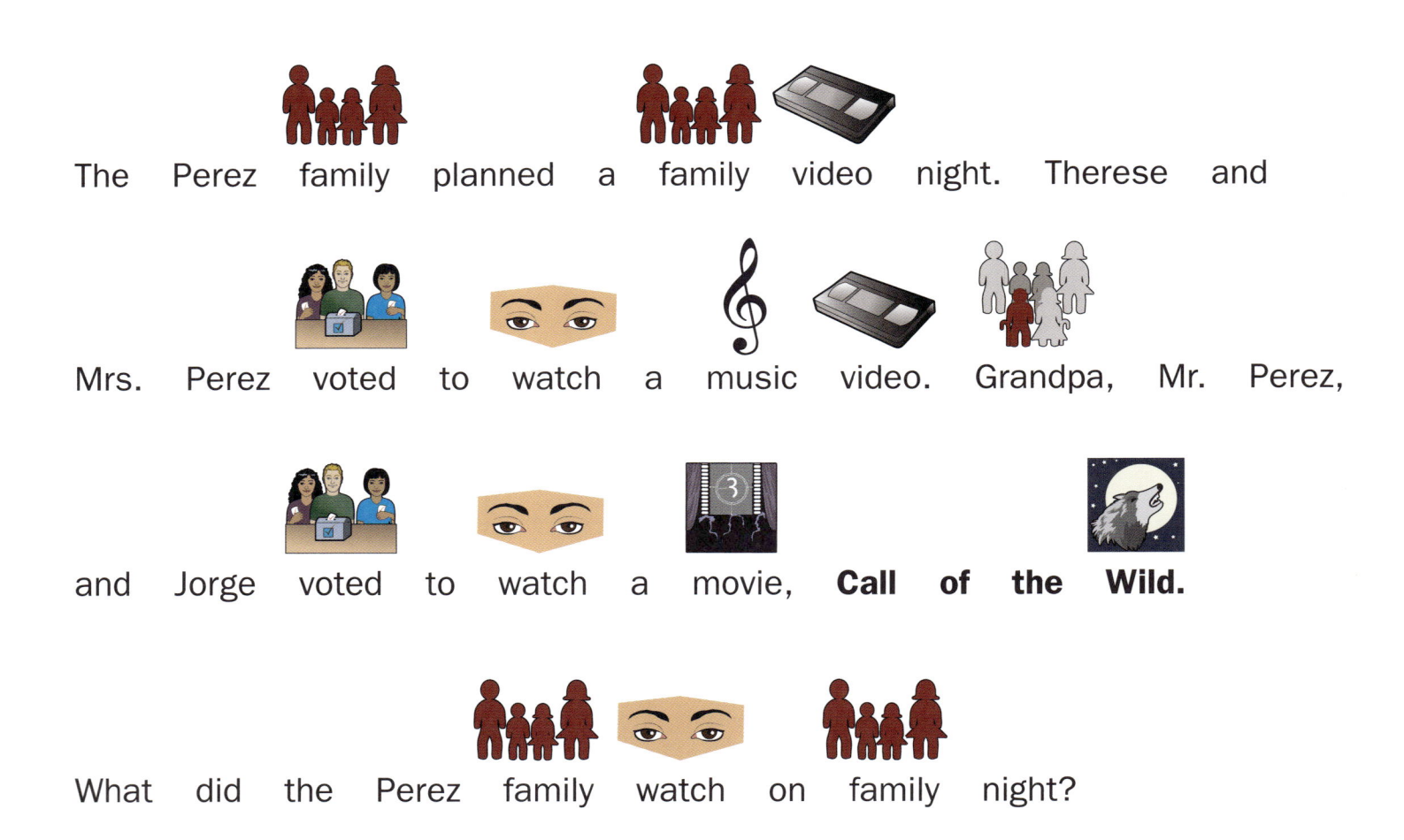

The Perez family planned a family video night. Therese and Mrs. Perez voted to watch a music video. Grandpa, Mr. Perez, and Jorge voted to watch a movie, **Call of the Wild.**

What did the Perez family watch on family night?

WHAT DO WE NEED TO FIND OUT? CHECK THE BOX. ☑

☐ 1. Where did the Perez family eat dinner?

☐ 2. What did the Perez family watch on family night?

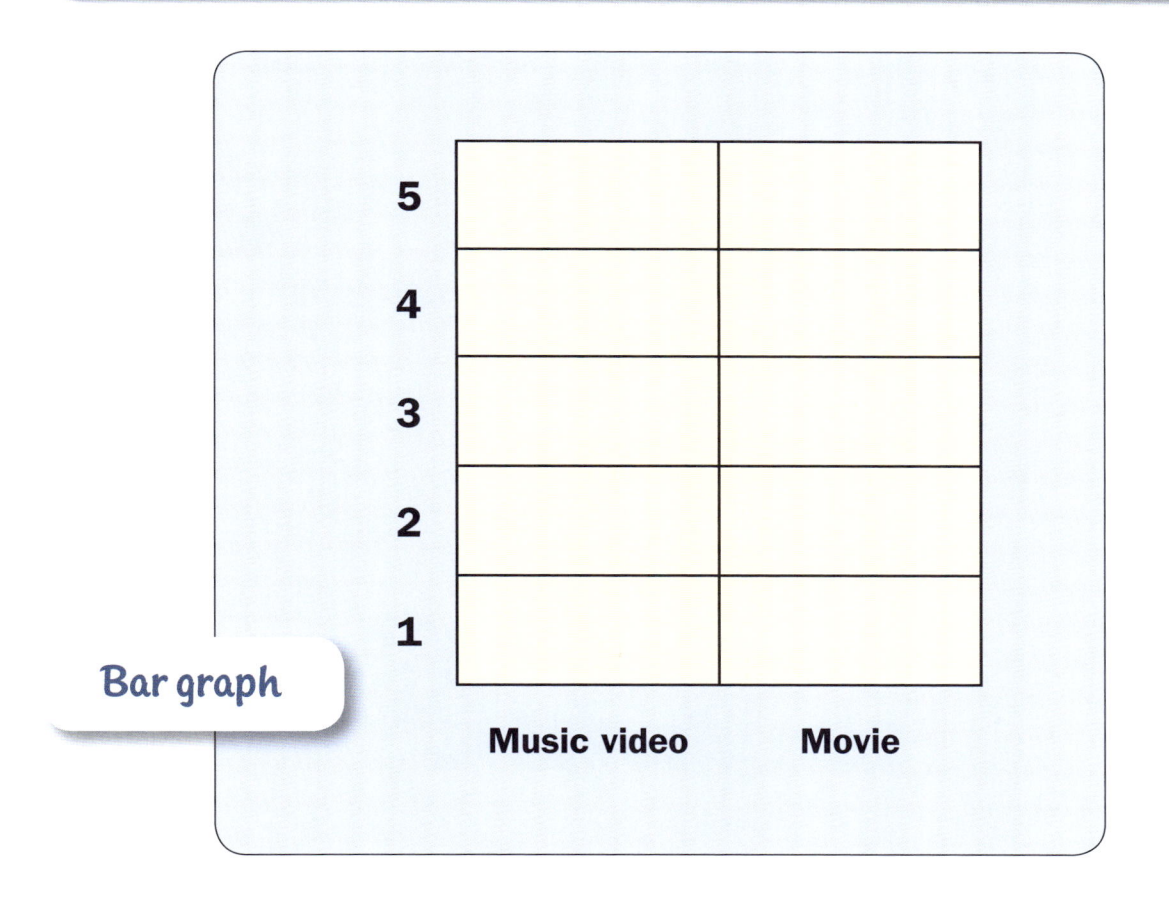

Bar graph

5
4
3
2
1

Music video   Movie

Video: _____

Unit 3: Data Analysis

87

Battle of the books

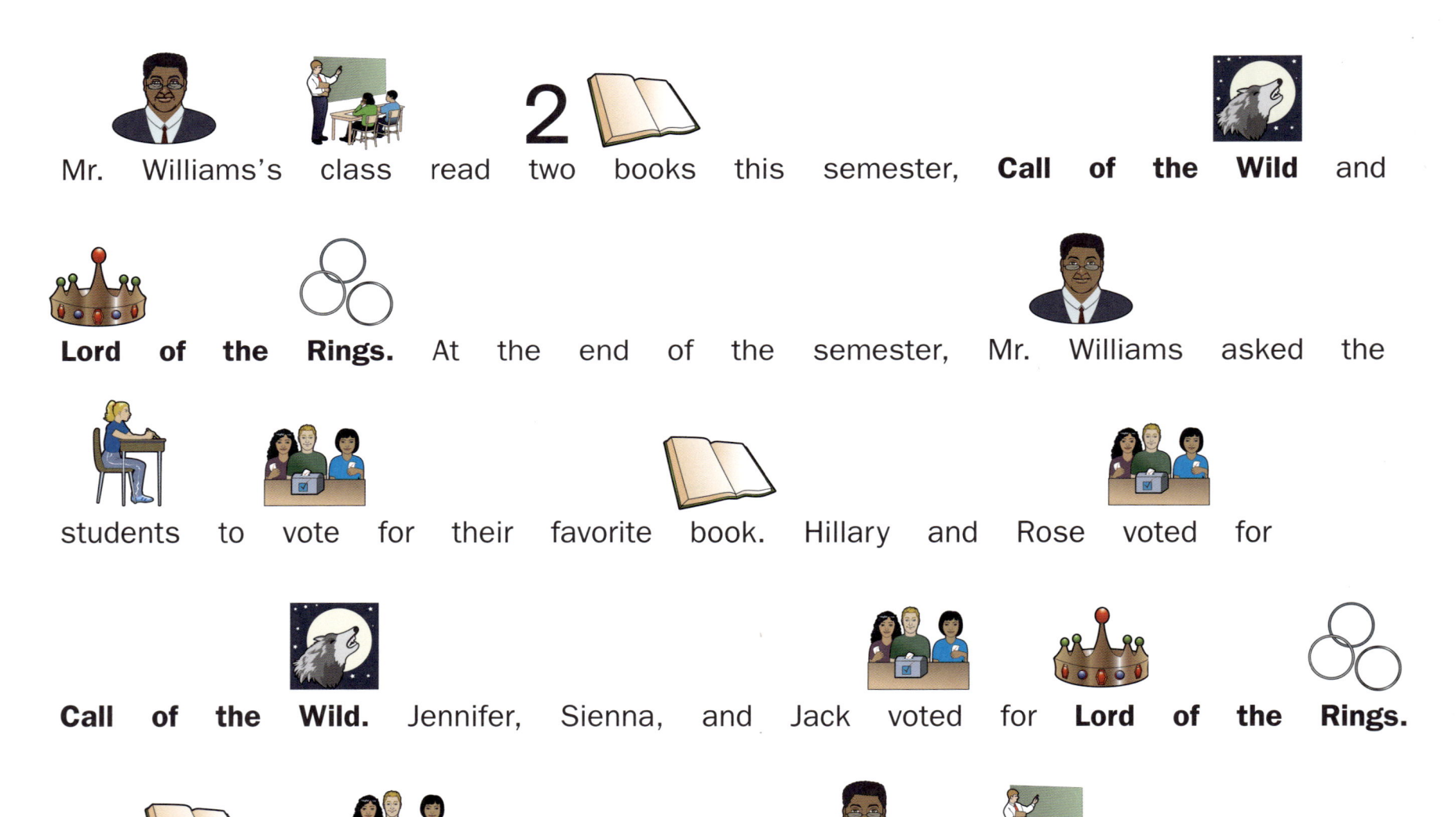

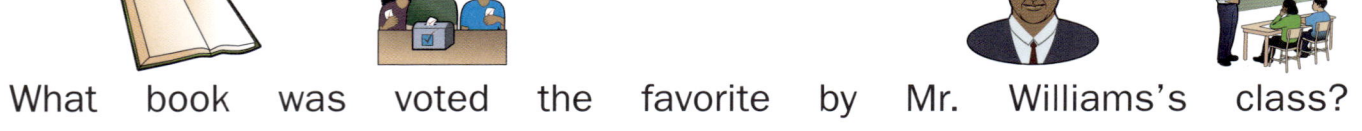

Mr. Williams's class read two books this semester, **Call of the Wild** and **Lord of the Rings.** At the end of the semester, Mr. Williams asked the students to vote for their favorite book. Hillary and Rose voted for **Call of the Wild.** Jennifer, Sienna, and Jack voted for **Lord of the Rings.**

What book was voted the favorite by Mr. Williams's class?

WHAT DO WE NEED TO FIND OUT? CHECK THE BOX. ☑

☐ 1. Where did Mr. Williams shop for books?

☐ 2. What book was voted the favorite by Mr. Williams's class?

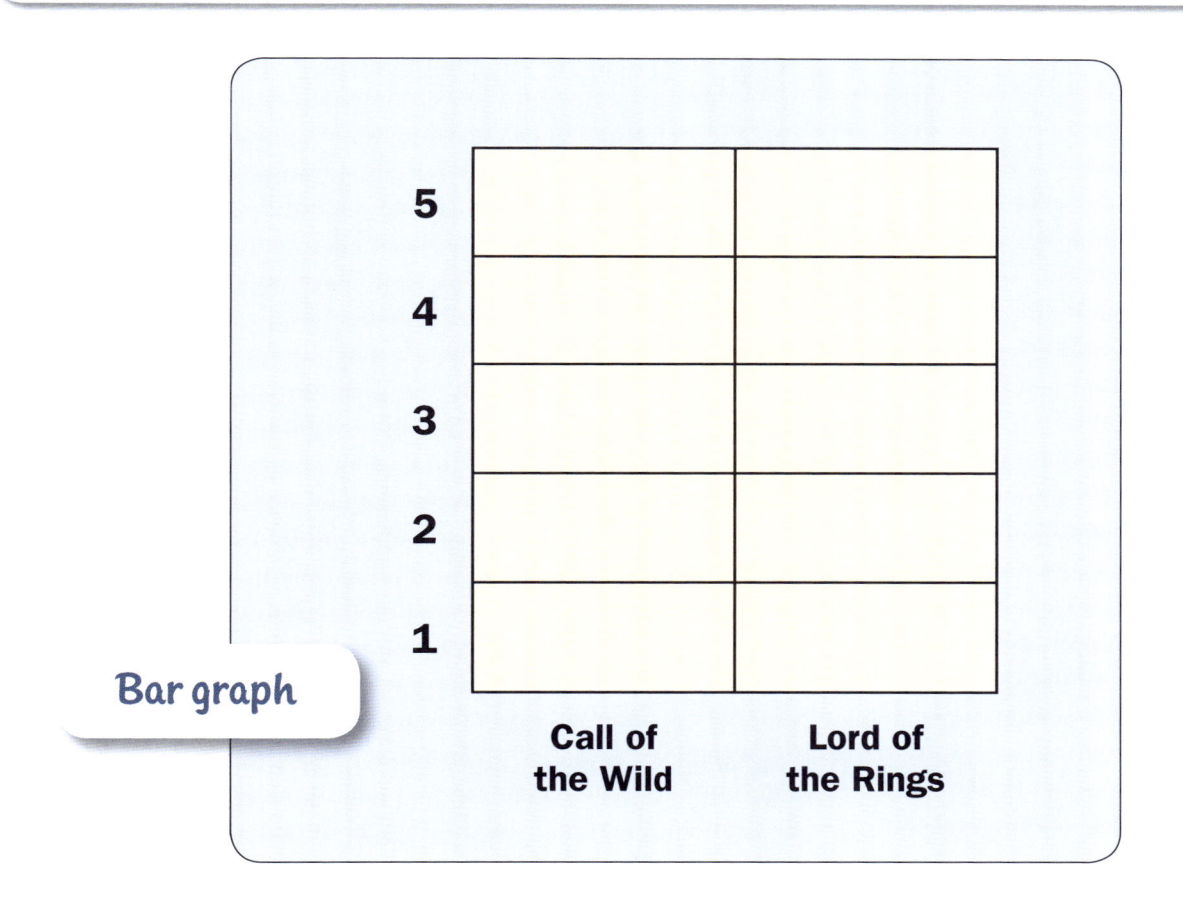

Bar graph

	Call of the Wild	Lord of the Rings
5		
4		
3		
2		
1		

Book: _____

Unit 3: Data Analysis

Student government elections

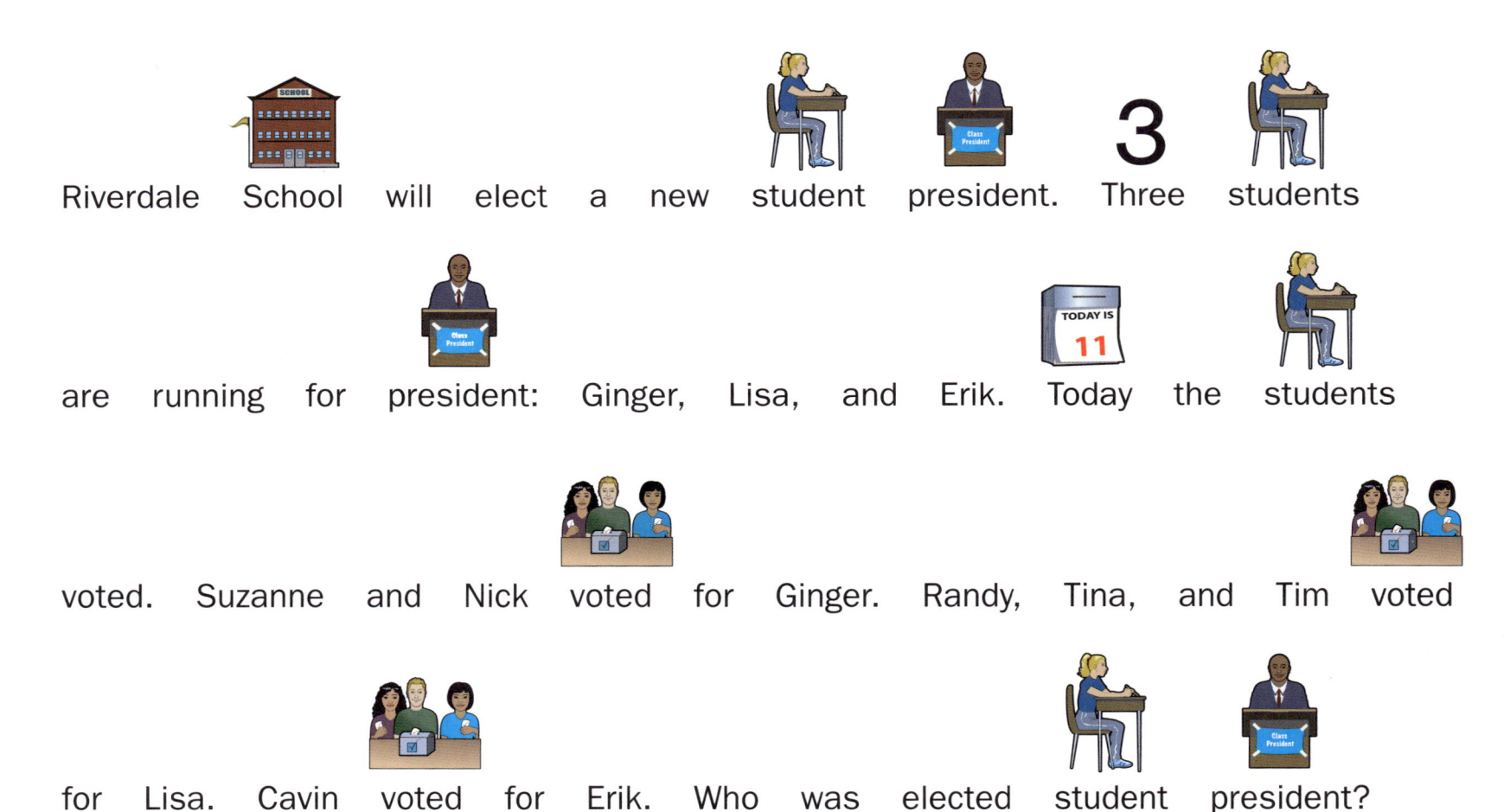

Riverdale School will elect a new student president. Three students are running for president: Ginger, Lisa, and Erik. Today the students voted. Suzanne and Nick voted for Ginger. Randy, Tina, and Tim voted for Lisa. Cavin voted for Erik. Who was elected student president?

WHAT DO WE NEED TO FIND OUT? CHECK THE BOX. ☑

☐ 1. Who was elected student president?

☐ 2. How many students attended Riverdale School?

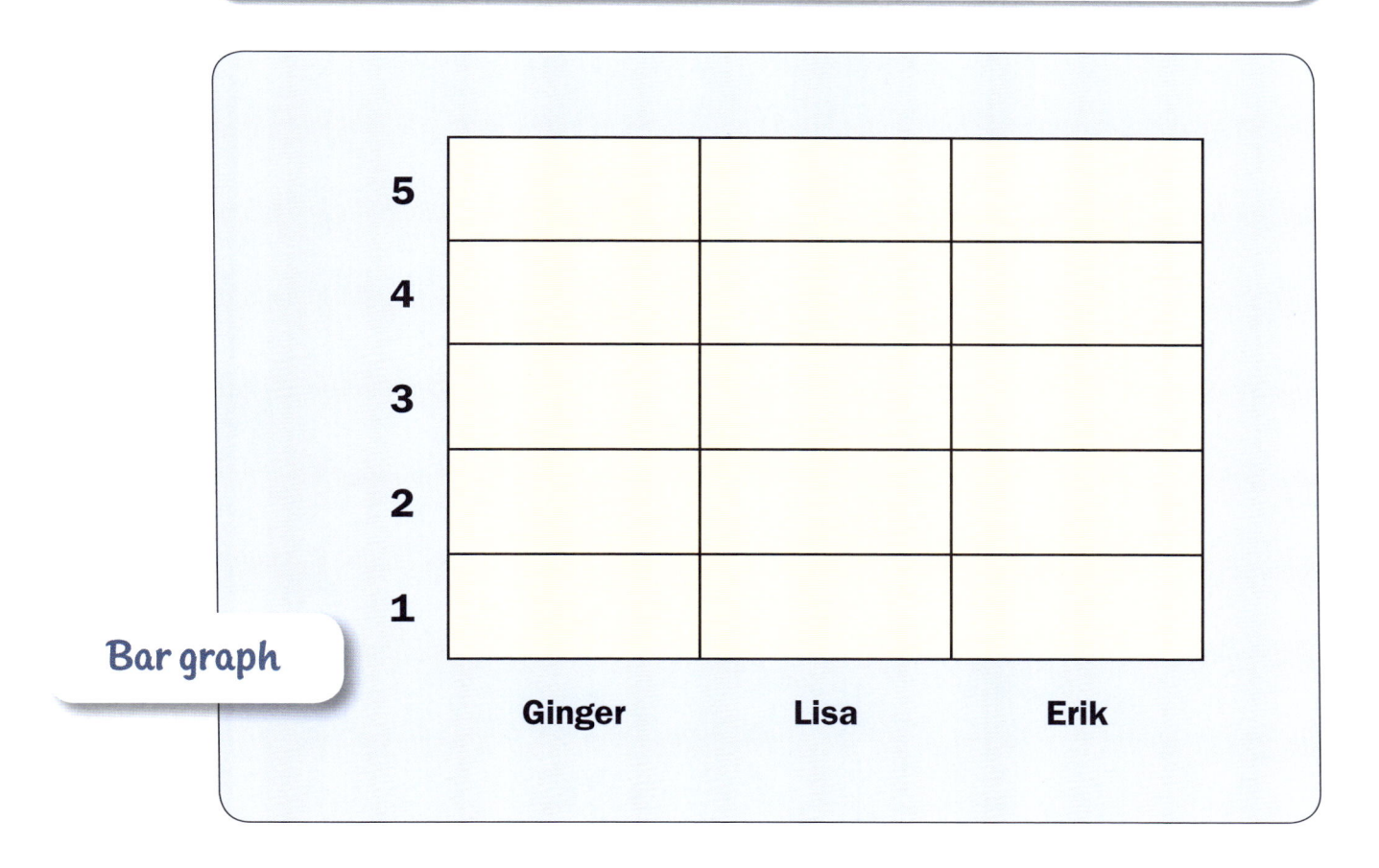

Bar graph

	Ginger	Lisa	Erik
5			
4			
3			
2			
1			

President: _____

Unit 3: Data Analysis

91

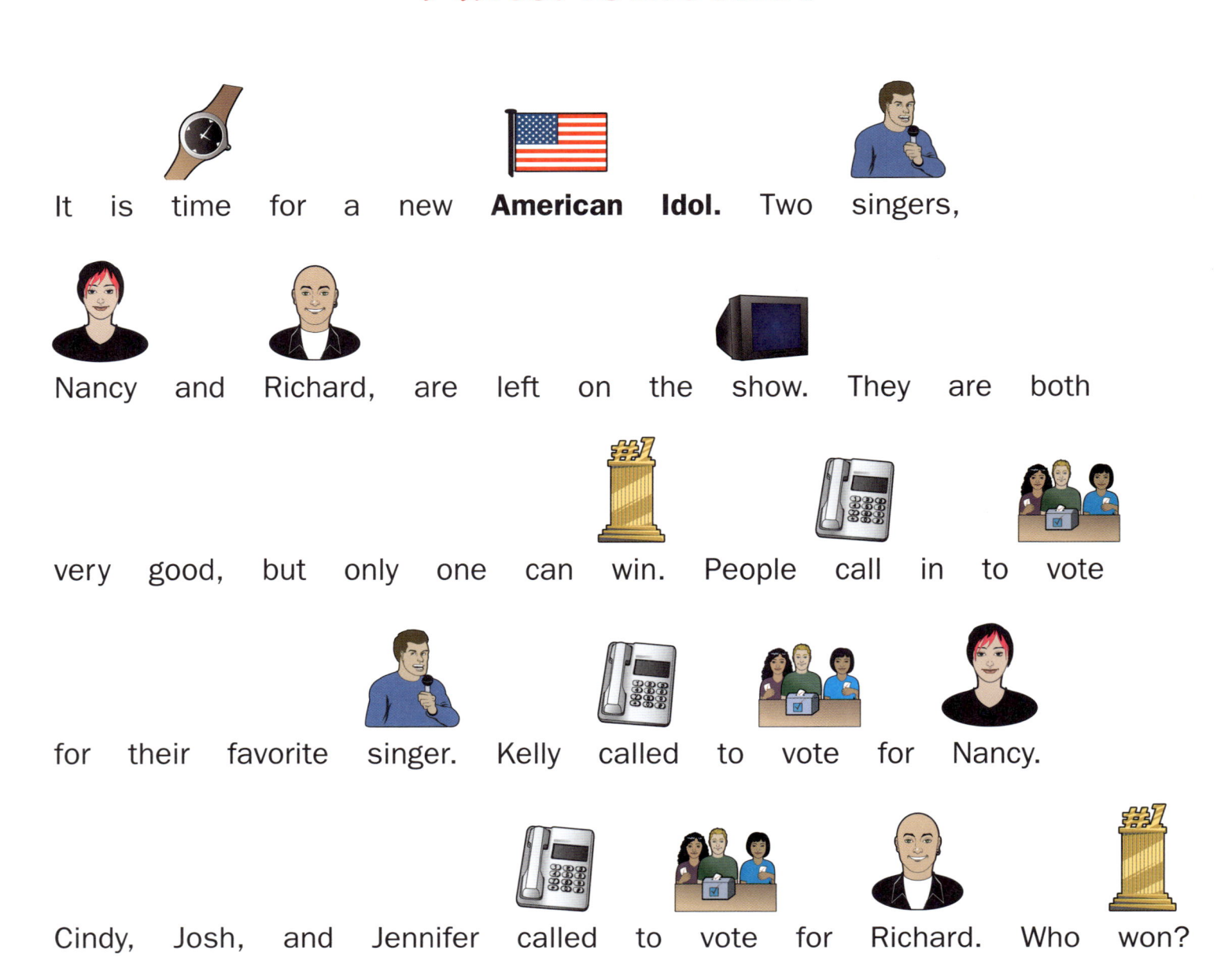

American Idol

It is time for a new **American Idol.** Two singers, Nancy and Richard, are left on the show. They are both very good, but only one can win. People call in to vote for their favorite singer. Kelly called to vote for Nancy. Cindy, Josh, and Jennifer called to vote for Richard. Who won?

WHAT DO WE NEED TO FIND OUT? CHECK THE BOX. ☑

☐ 1. Why do people watch **American Idol?**

☐ 2. Who won?

	Nancy	Richard
5		
4		
3		
2		
1		

Bar graph

Singer: _____

Student of the Month

Mrs. Kimbel's class will send one student to a party    for Student of the Month. Mrs. Kimbel picked Regina and   Smitty because they did all of their homework. Tim, Katie,   Cliff, and Vanessa voted for Regina. Samantha, Ron, and Kayla voted for Smitty. Who was Student of the Month?

WHAT DO WE NEED TO FIND OUT? CHECK THE BOX. ☑

☐ 1. How much homework did Regina and Smitty have?

☐ 2. Who was Student of the Month?

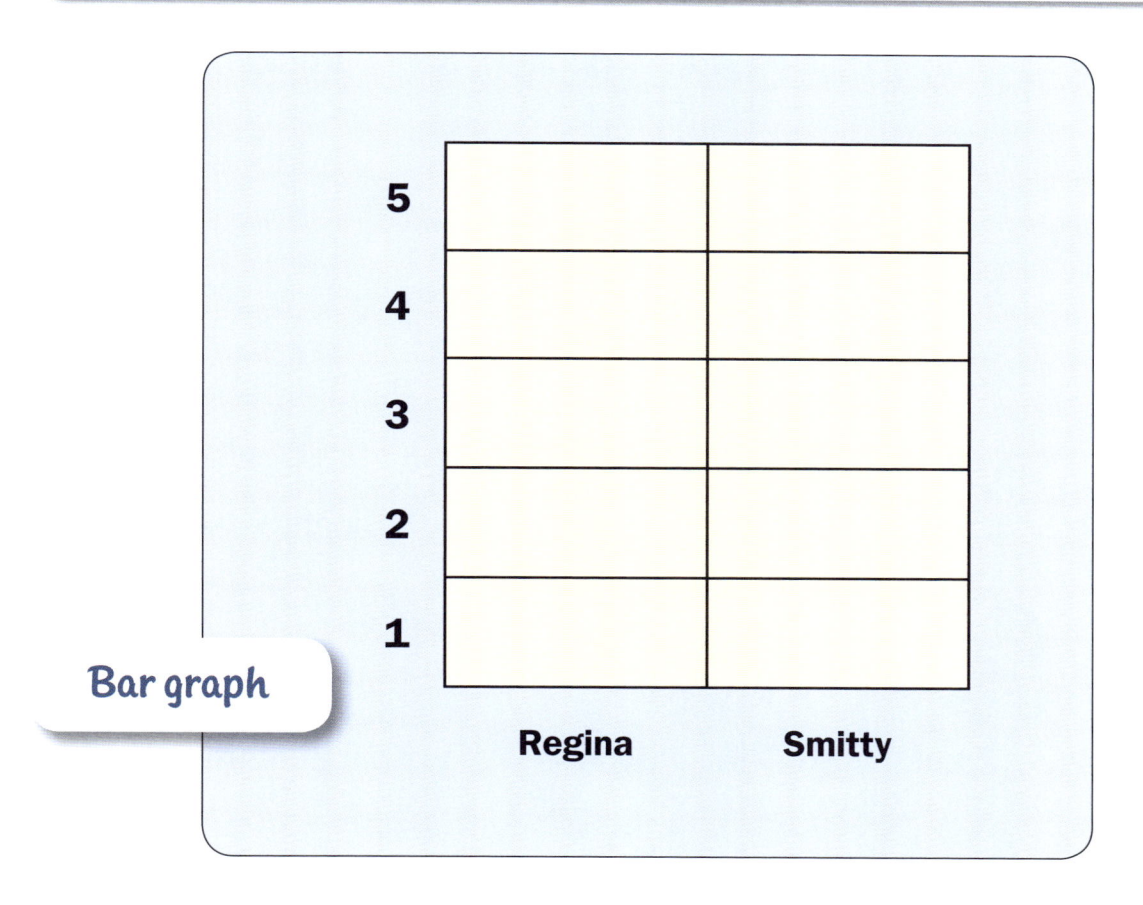

Bar graph

5
4
3
2
1

Regina   Smitty

Student: _____

Unit 3: Data Analysis

95

A new president

Mr. Taylor and Ms. Black are running for president of the    Parent Teacher Student Association. Mr. Lee and Mrs. Perez   voted for Mr. Taylor. Grandpa Perez, Ms. Singh, and Mr. Kelly voted for Ms. Black. So far, who is winning the election?

WHAT DO WE NEED TO FIND OUT? CHECK THE BOX. ☑

☐ 1. How much did the school spend on the election?

☐ 2. So far, who is winning the election?

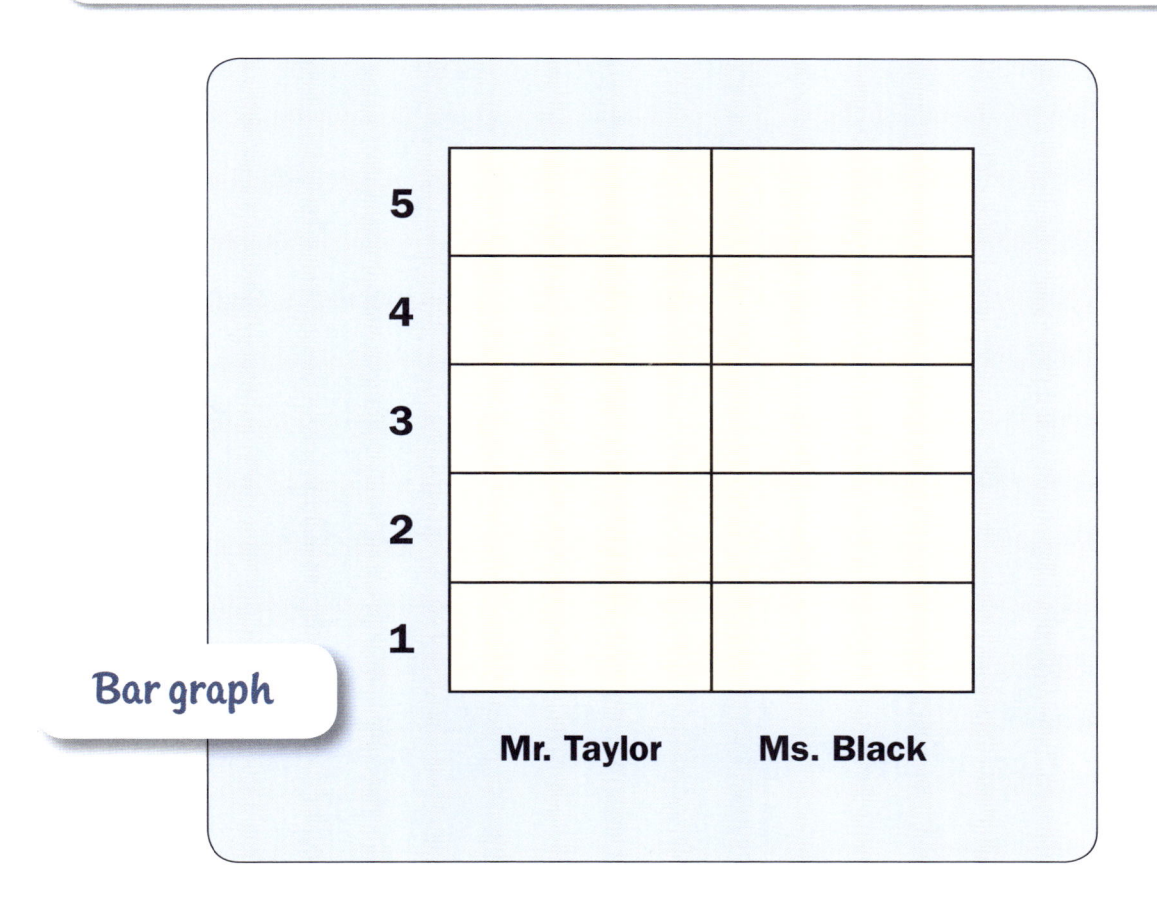

Bar graph

5
4
3
2
1

Mr. Taylor   Ms. Black

President: _____

Unit 3: Data Analysis

97

Teacher of the Month

The students at Riverdale school are voting for   Teacher of the Month. Randy, Tina, Irene, and Tim    voted for Mr. Fig. Gavin, Therese, and Lisa voted   for Ms. Black. So far, who is winning the election?

WHAT DO WE NEED TO FIND OUT? CHECK THE BOX. ✓

☐ 1. So far, who is winning the election?

☐ 2. How many teachers work at Riverdale School?

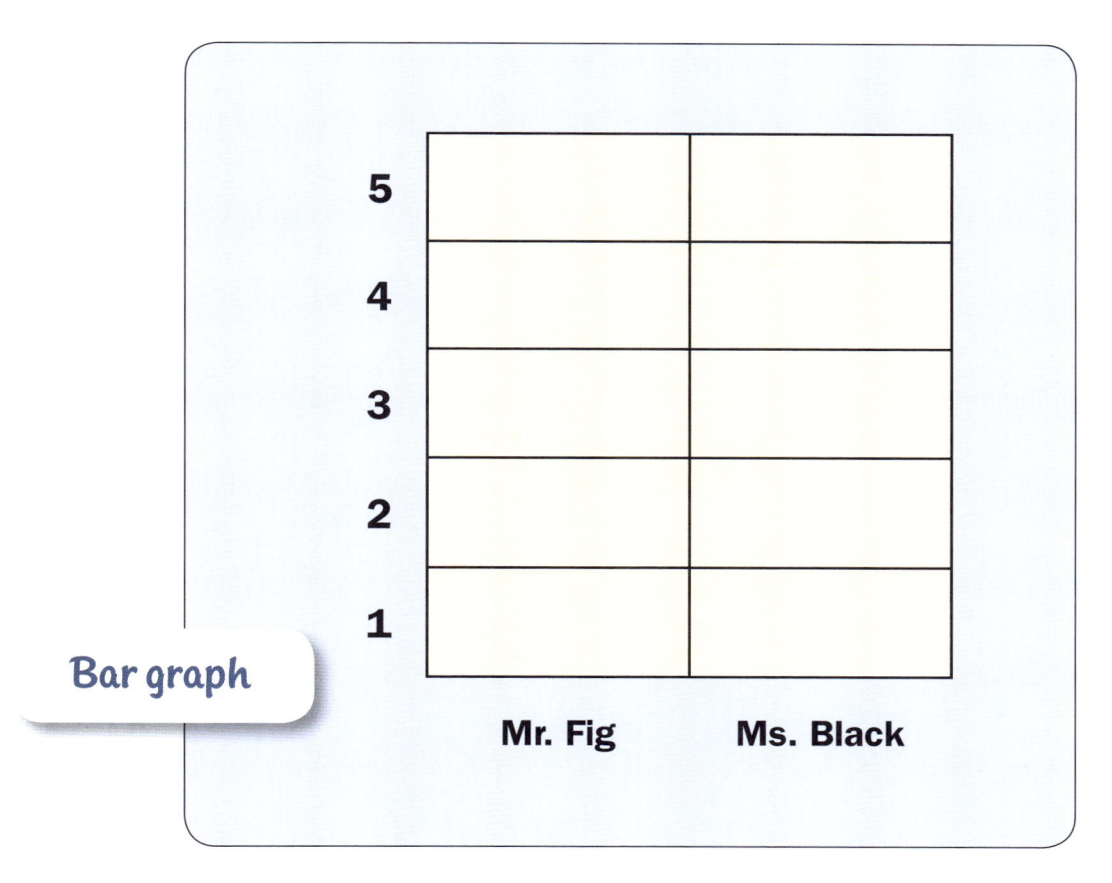

Bar graph

5
4
3
2
1

Mr. Fig   Ms. Black

Teacher: _____

Unit 3: Data Analysis

Sisters share an apartment

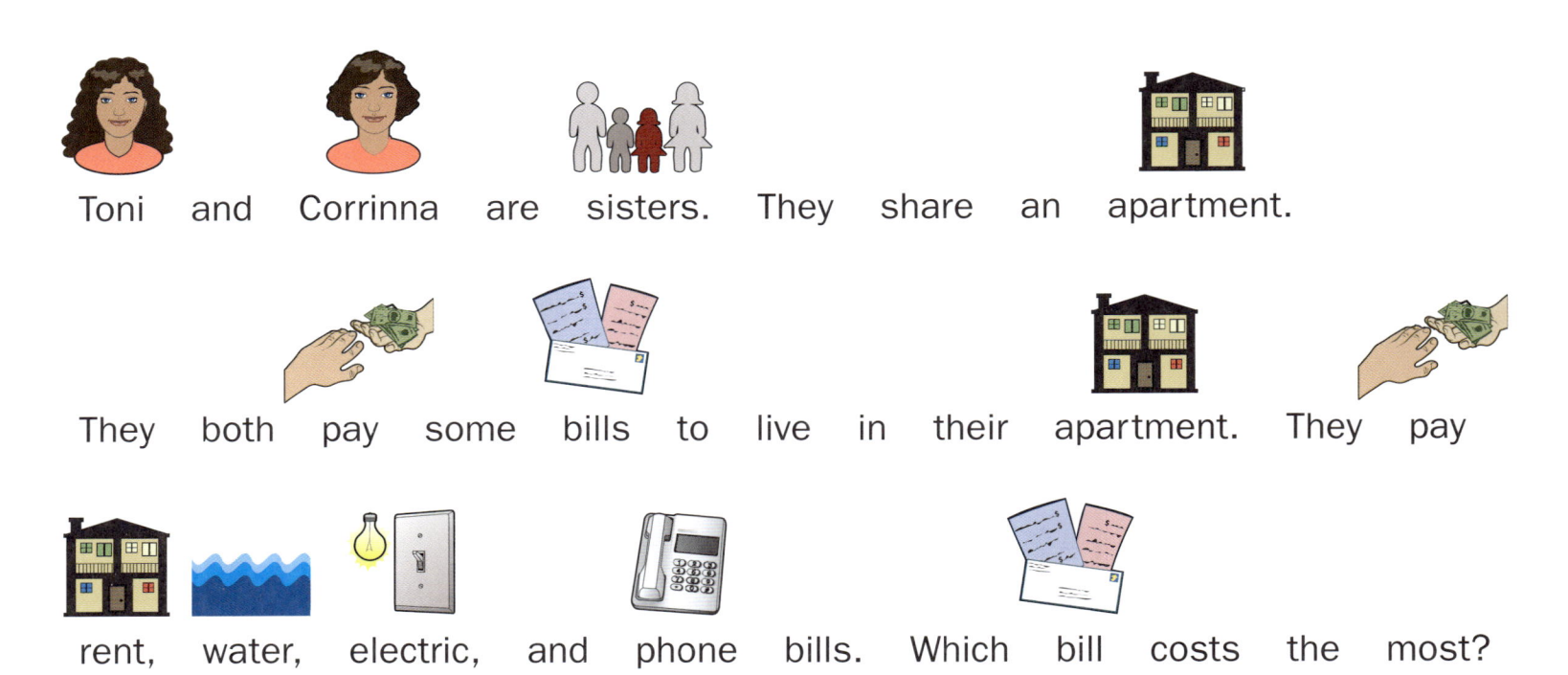

Toni and Corrinna are sisters. They share an apartment. They both pay some bills to live in their apartment. They pay rent, water, electric, and phone bills. Which bill costs the most?

WHAT DO WE NEED TO FIND OUT? CHECK THE BOX. ☑

- [ ] 1. Where do the sisters live?
- [ ] 2. Which bill costs the most?

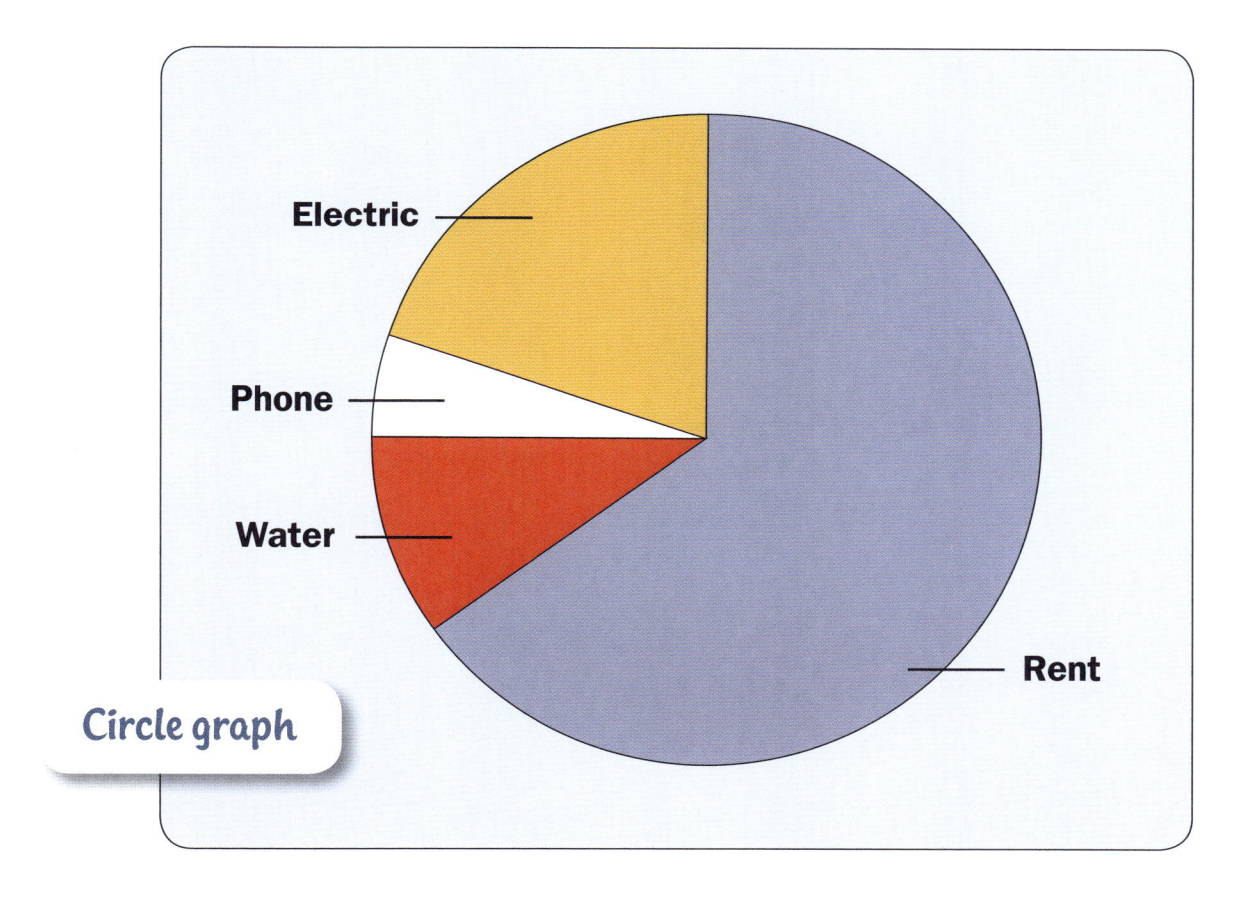

Circle graph

- Electric
- Phone
- Water
- Rent

Bill: _____

Unit 3: Data Analysis

School friends share expenses

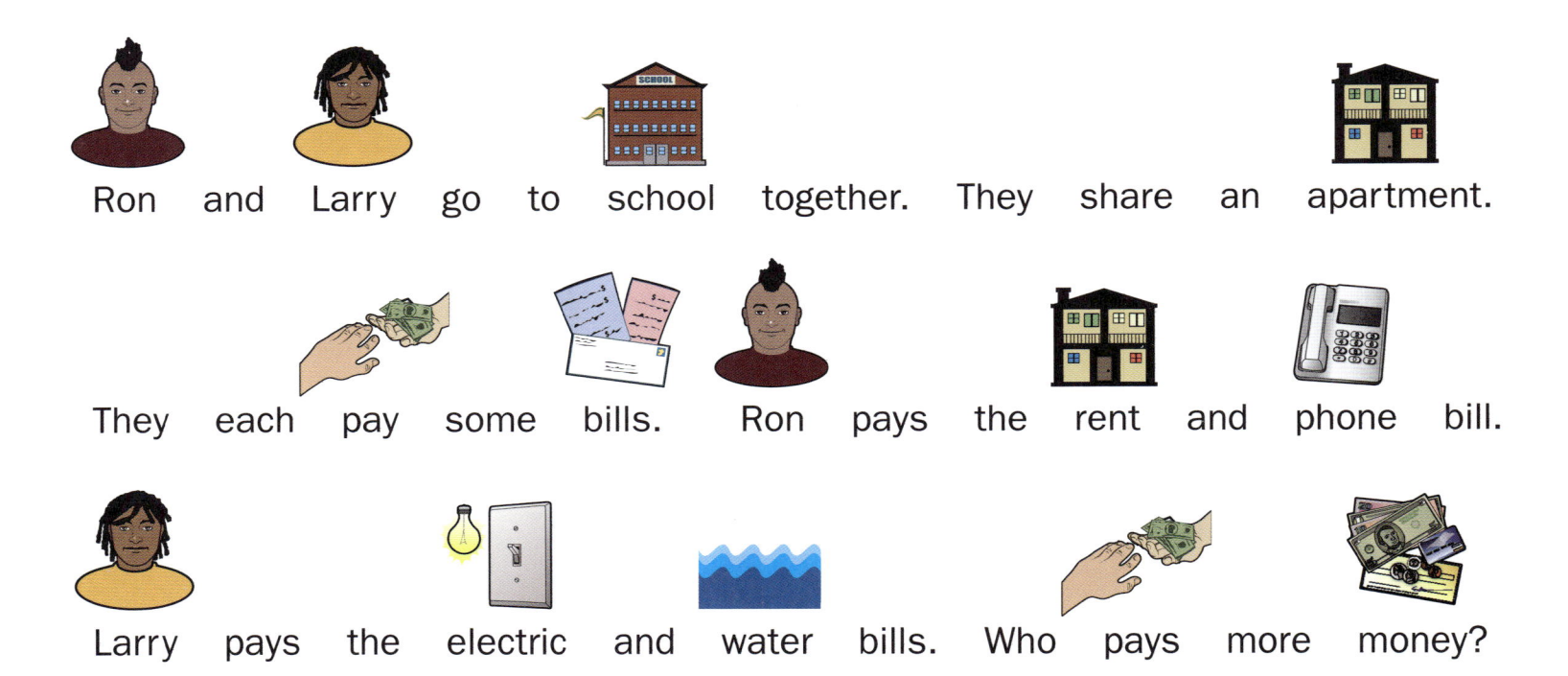

Ron and Larry go to school together. They share an apartment.

They each pay some bills. Ron pays the rent and phone bill.

Larry pays the electric and water bills. Who pays more money?

WHAT DO WE NEED TO FIND OUT? CHECK THE BOX. ☑

☐ 1. Who pays more money?

☐ 2. What streets do Ron and Larry travel to go to school?

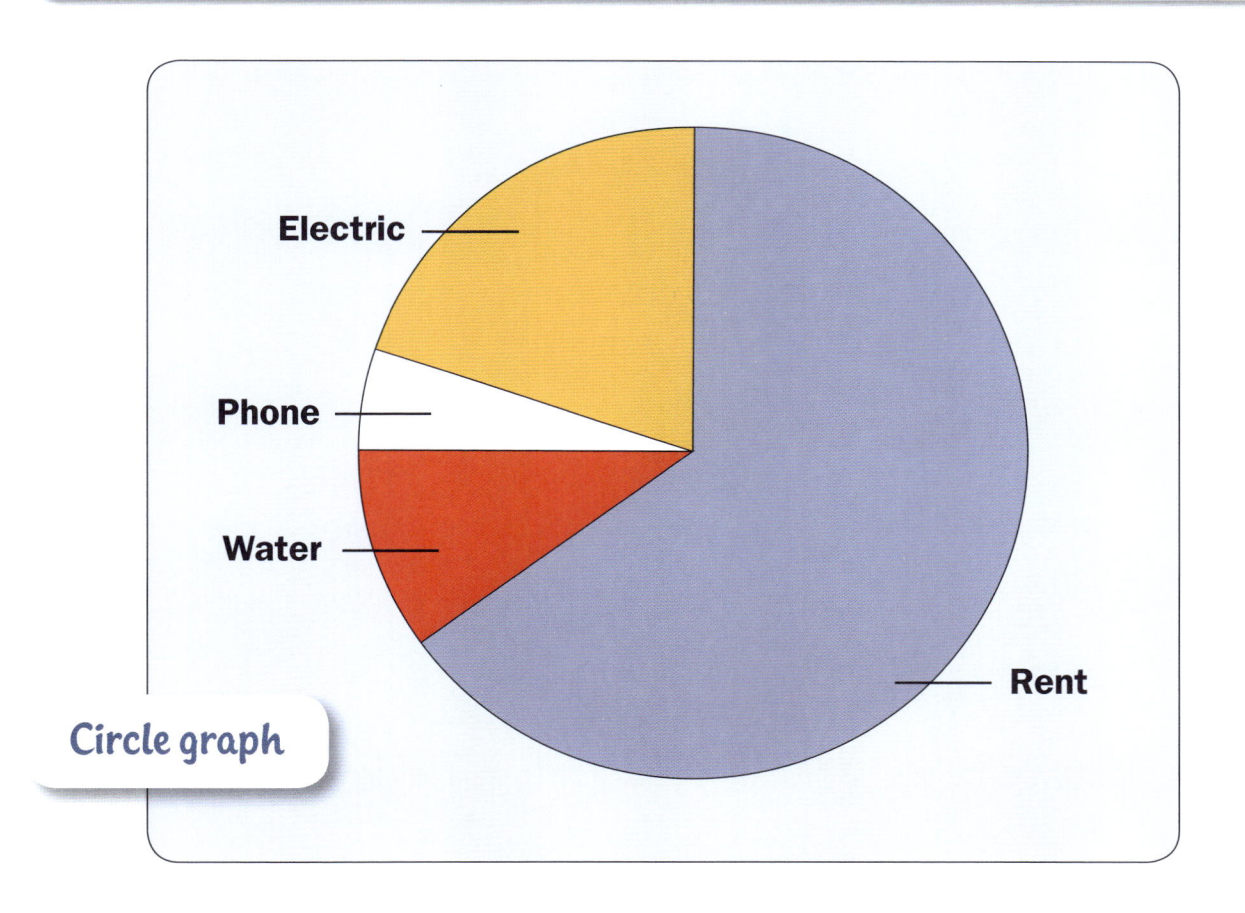

Electric

Phone

Water

Rent

Circle graph

Student: _____

Pavel's plans

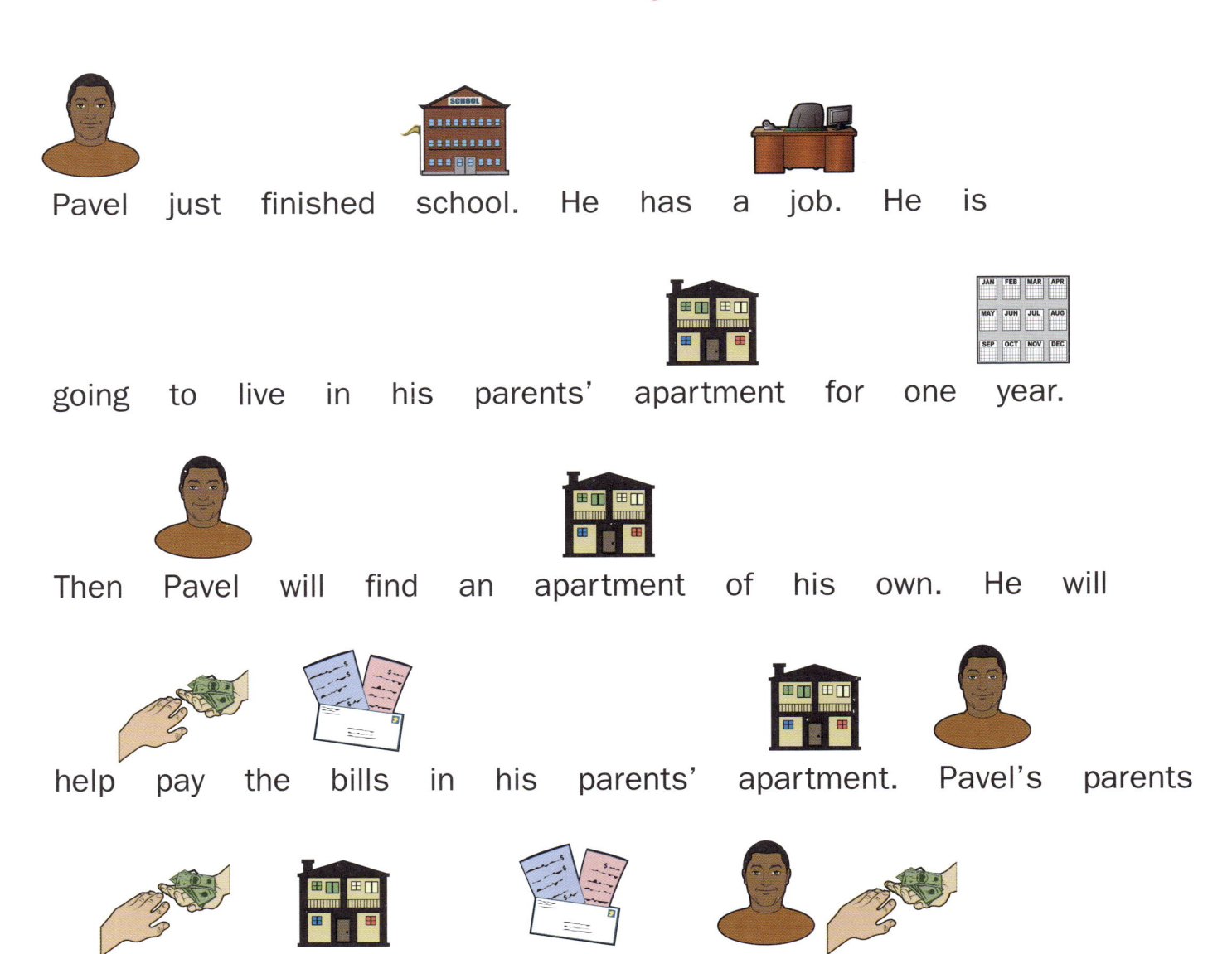

Pavel just finished school. He has a job. He is going to live in his parents' apartment for one year.

Then Pavel will find an apartment of his own. He will help pay the bills in his parents' apartment. Pavel's parents will pay the rent. Which bills will Pavel pay?

WHAT DO WE NEED TO FIND OUT? CHECK THE BOX. ☑

☐ 1. Which bills will Pavel pay?

☐ 2. How long will Pavel live in his parents' apartment?

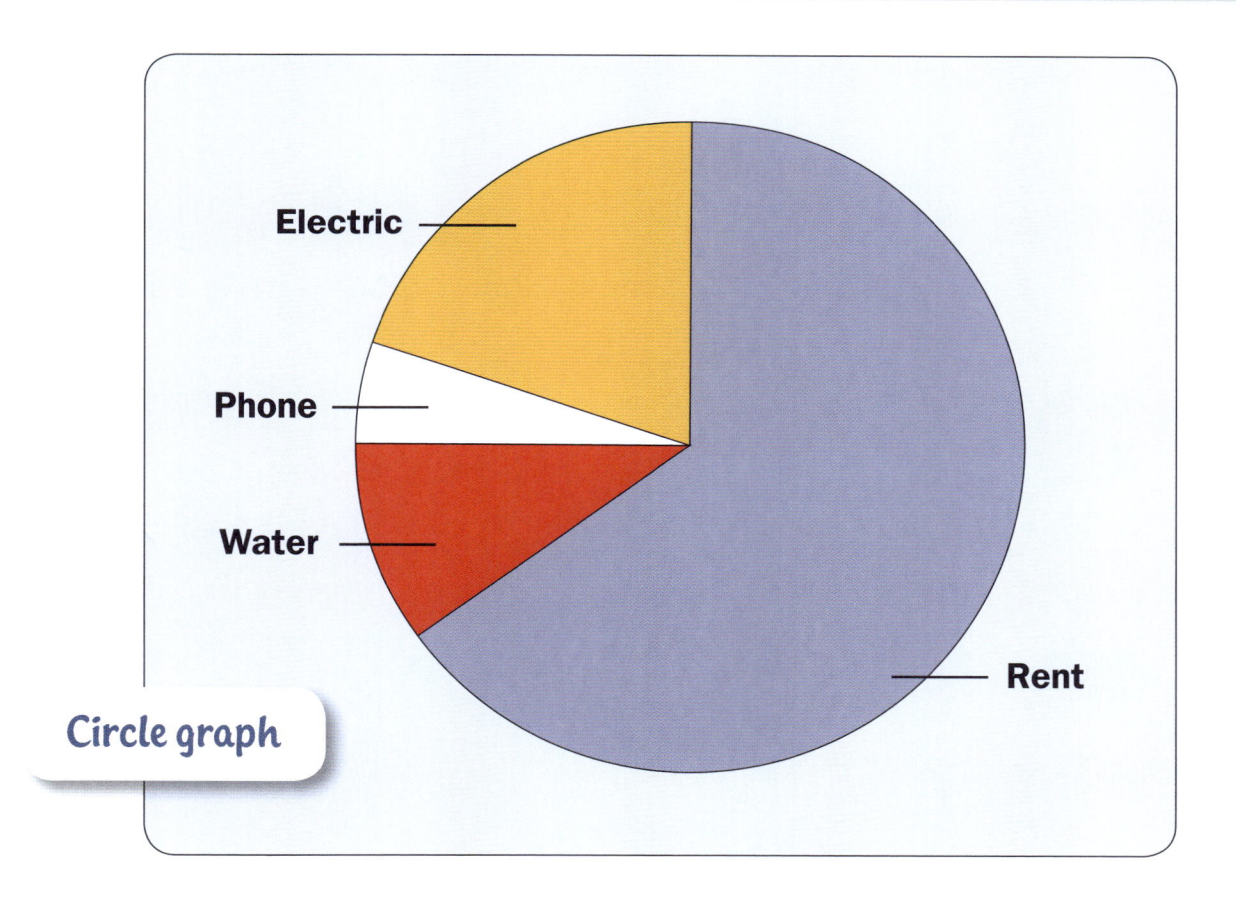

Circle graph

- Electric
- Phone
- Water
- Rent

Bills: _____

Unit 3: Data Analysis

105

Randy's day

Randy's mom asked him how he spent his school day.

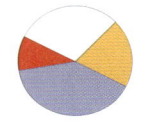

Randy made a circle graph to show his mom what he does     during his school day. Randy spent time in technology class, 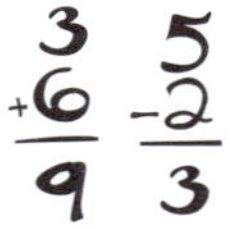    math class, PE, and English. Where did Randy spend   the least amount of time in his school day?

WHAT DO WE NEED TO FIND OUT? CHECK THE BOX. ☑

☐ 1. Where did Randy spend the least amount of time in his school day?

☐ 2. What is Randy's favorite food?

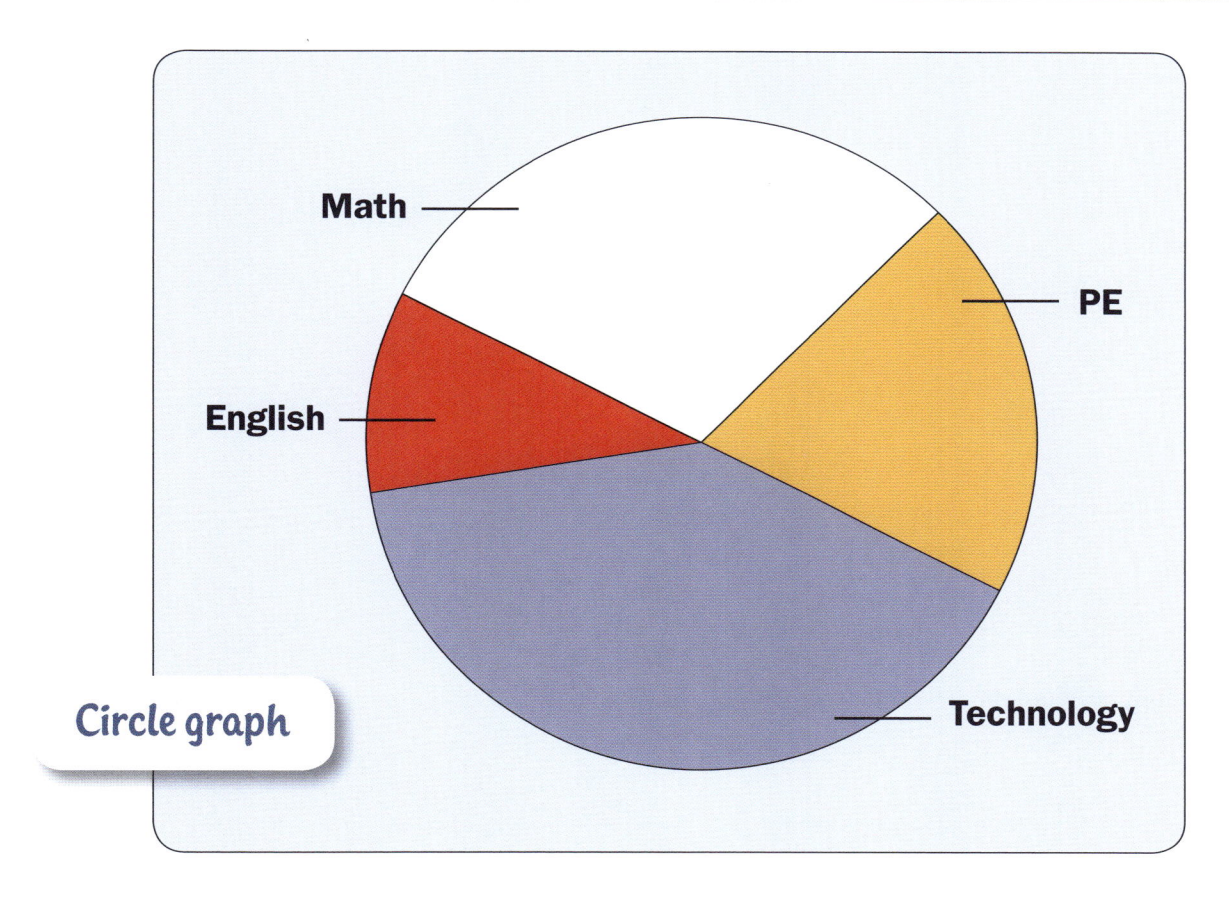

Circle graph

- Math
- PE
- English
- Technology

Class: _____

Unit 3: Data Analysis

Favorite classes

Imani and his friends are talking about their school day.

They talk about their favorite classes. Imani says he spends most of his time at school in his favorite class.

What is Imani's favorite class?

WHAT DO WE NEED TO FIND OUT? CHECK THE BOX. ☑

☐ 1. What is Imani's favorite class?

☐ 2. Where does Imani shop for books?

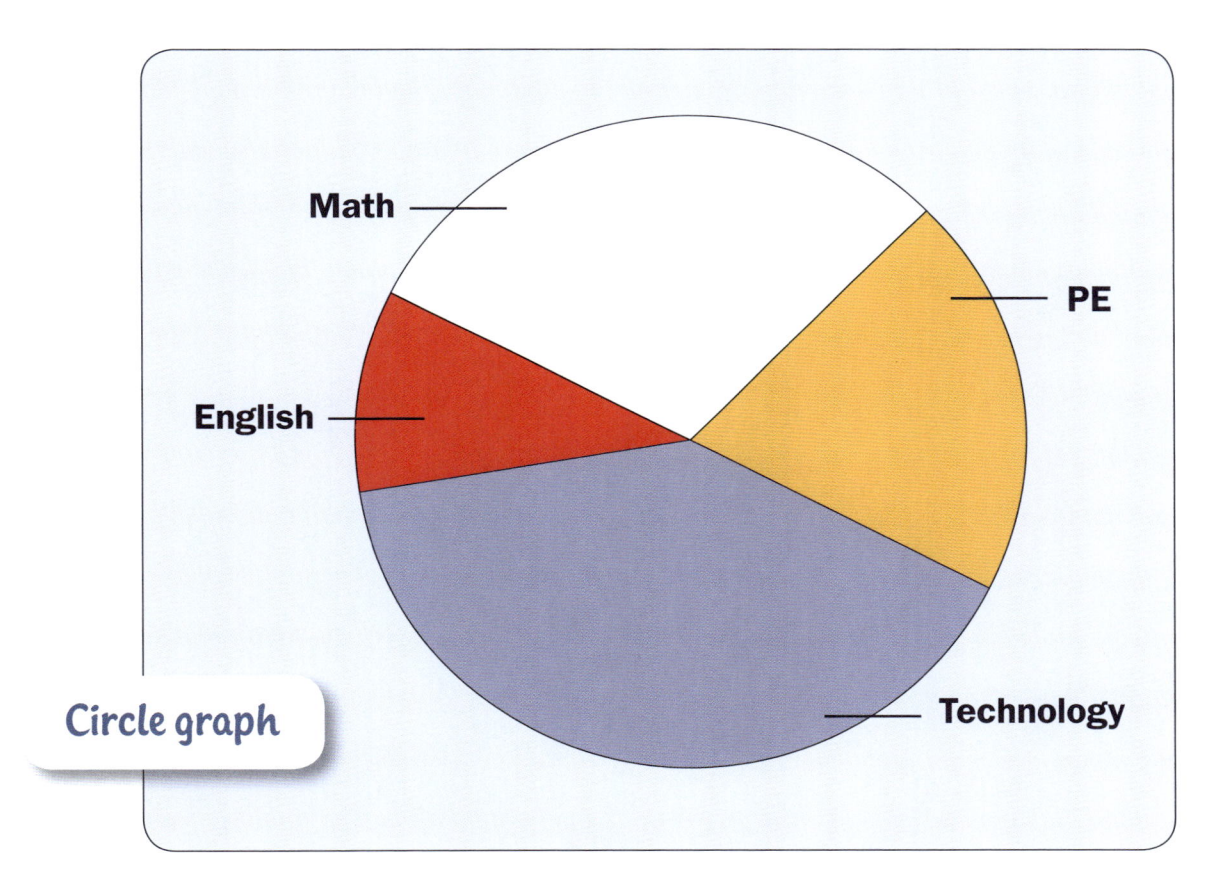

Circle graph

- Math
- PE
- English
- Technology

Class: _____

Unit 3: Data Analysis

109

A class to practice writing

Kris has a meeting with her guidance counselor.

Kris needs to improve her grades in writing next semester.

Her counselor said that Kris can practice writing  in the class where she spends most of her     school day. In what class will Kris practice writing?

WHAT DO WE NEED TO FIND OUT? CHECK THE BOX. ☑

☐ 1. In what class will Kris practice writing?

☐ 2. Who is Kris's guidance counselor?

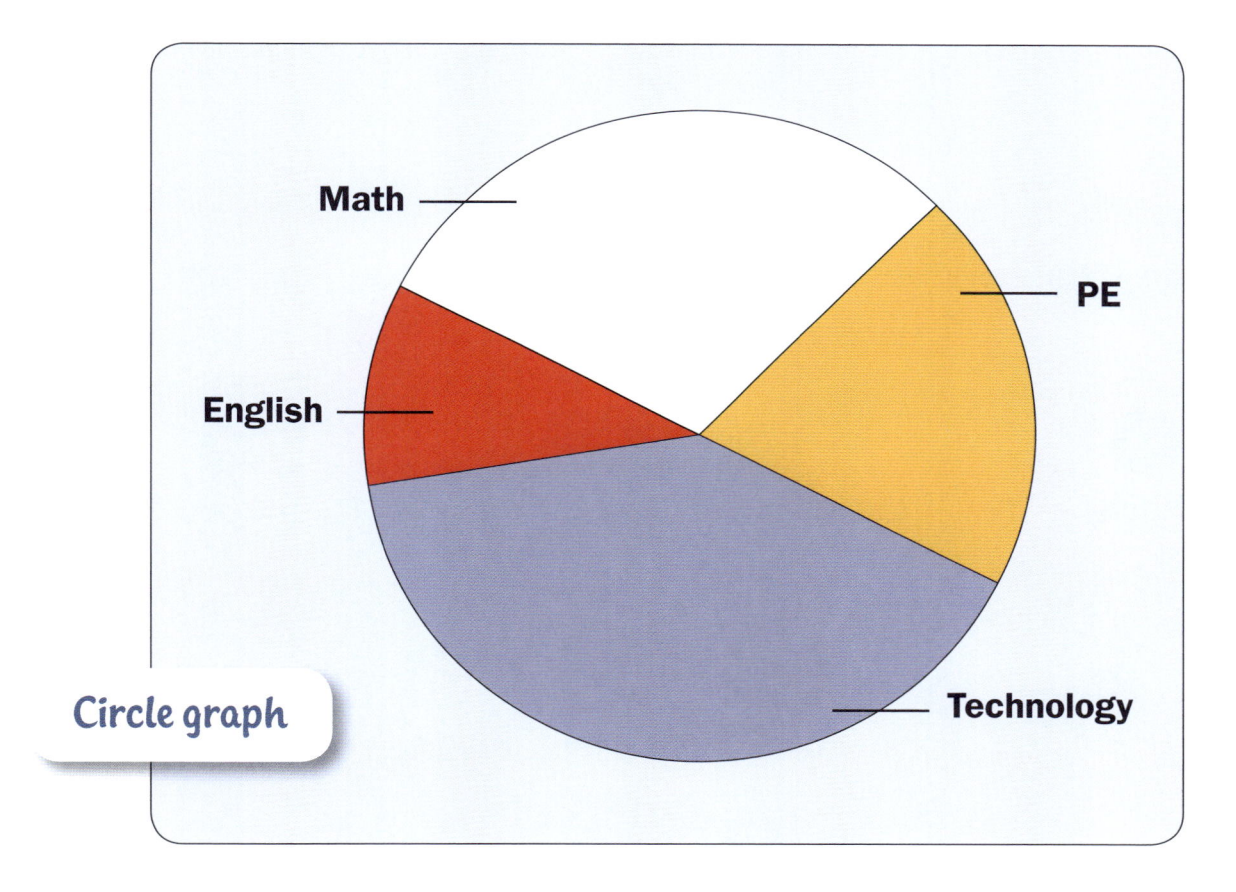

Circle graph

- Math
- PE
- English
- Technology

Class: _____

Unit 3: Data Analysis

The Cruz family rents a movie

The Cruz family will rent a movie. Carla wants to rent    **Lord of the Rings.** Her brother, Miguel, wants to rent    **Men in Black.** They decided to vote on what movie to rent. Dad and Grandpa chose **Lord of the Rings.** Aunt Rosa chose **Men in Black.** What movie will the Cruz family rent?

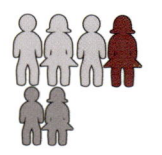

WHAT DO WE NEED TO FIND OUT? CHECK THE BOX. ☑

☐ 1. What movie will the Cruz family rent?

☐ 2. Where is the movie store?

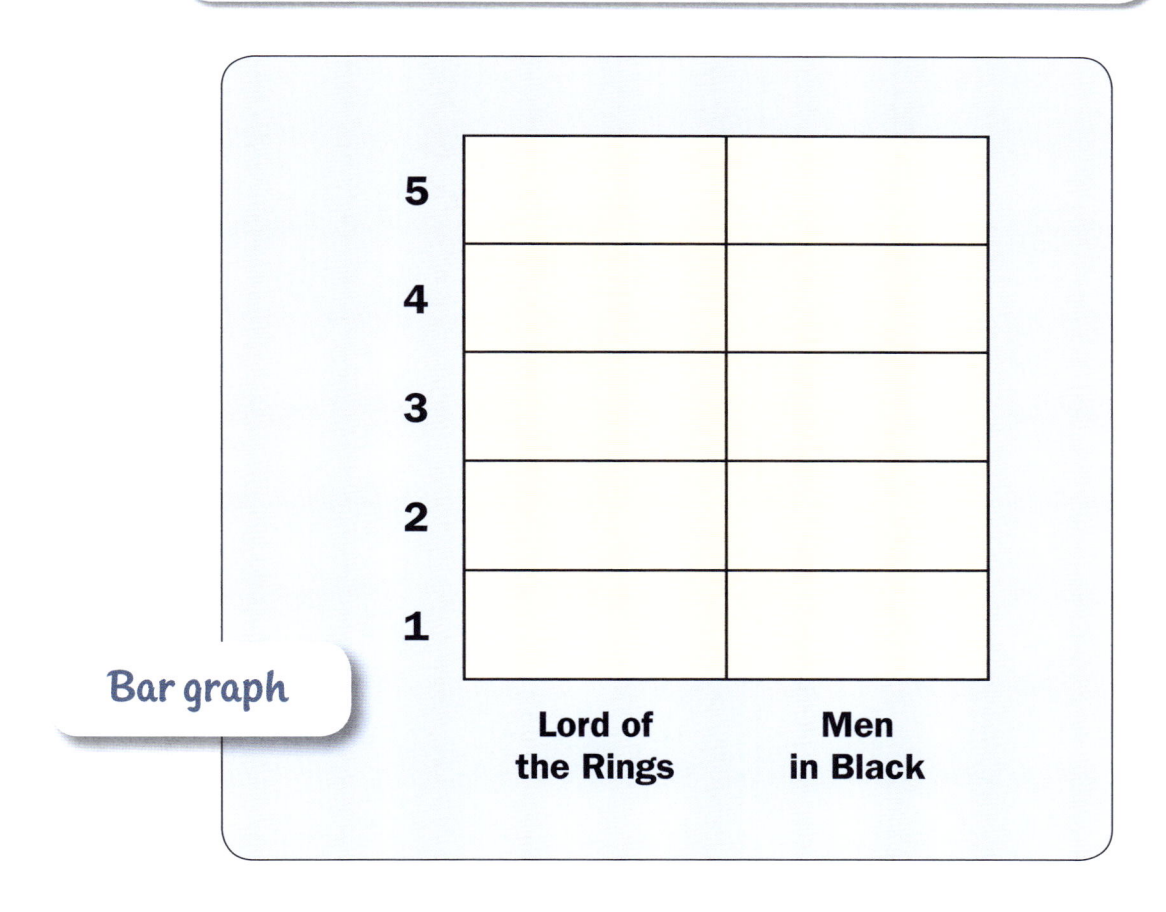

Bar graph

5
4
3
2
1

Lord of the Rings | Men in Black

Movie: _____

Unit 3: Data Analysis

113

UNIT 4 MEASUREMENT

At the matinee

Tao and his little sister, Li-Li, went to a matinee.

A matinee is a movie shown in the afternoon.

Tickets cost less for a matinee. Tao had some dollar bills     in his wallet to pay for the tickets. Altogether, the tickets cost $6.50. How much money did Tao give the cashier?

WHAT DO WE NEED TO FIND OUT? CHECK THE BOX. ☑

☐ 1. What movie did Tao and Li-Li see?

☐ 2. How much money did Tao give the cashier?

1 2 3 4 5 6 7 8 9 10

Next dollar line

$ _____

Unit 4: Measurement

Popcorn at the movies

Morgan and her friends went to see a movie.

Morgan has a gift card to pay for her ticket.

She put some dollar bills in her wallet to buy popcorn.

Morgan chose the large bucket of popcorn for $3.30.

How much money did Morgan give the cashier for her popcorn?

WHAT DO WE NEED TO FIND OUT? CHECK THE BOX. ☑

☐ 1. Who gave Morgan the gift card?

☐ 2. How much money did Morgan give the cashier for the popcorn?

1 2 3 4 5 6 7 8 9 10

Next dollar line

$ _____

Unit 4: Measurement

Casey's family at the drive-in

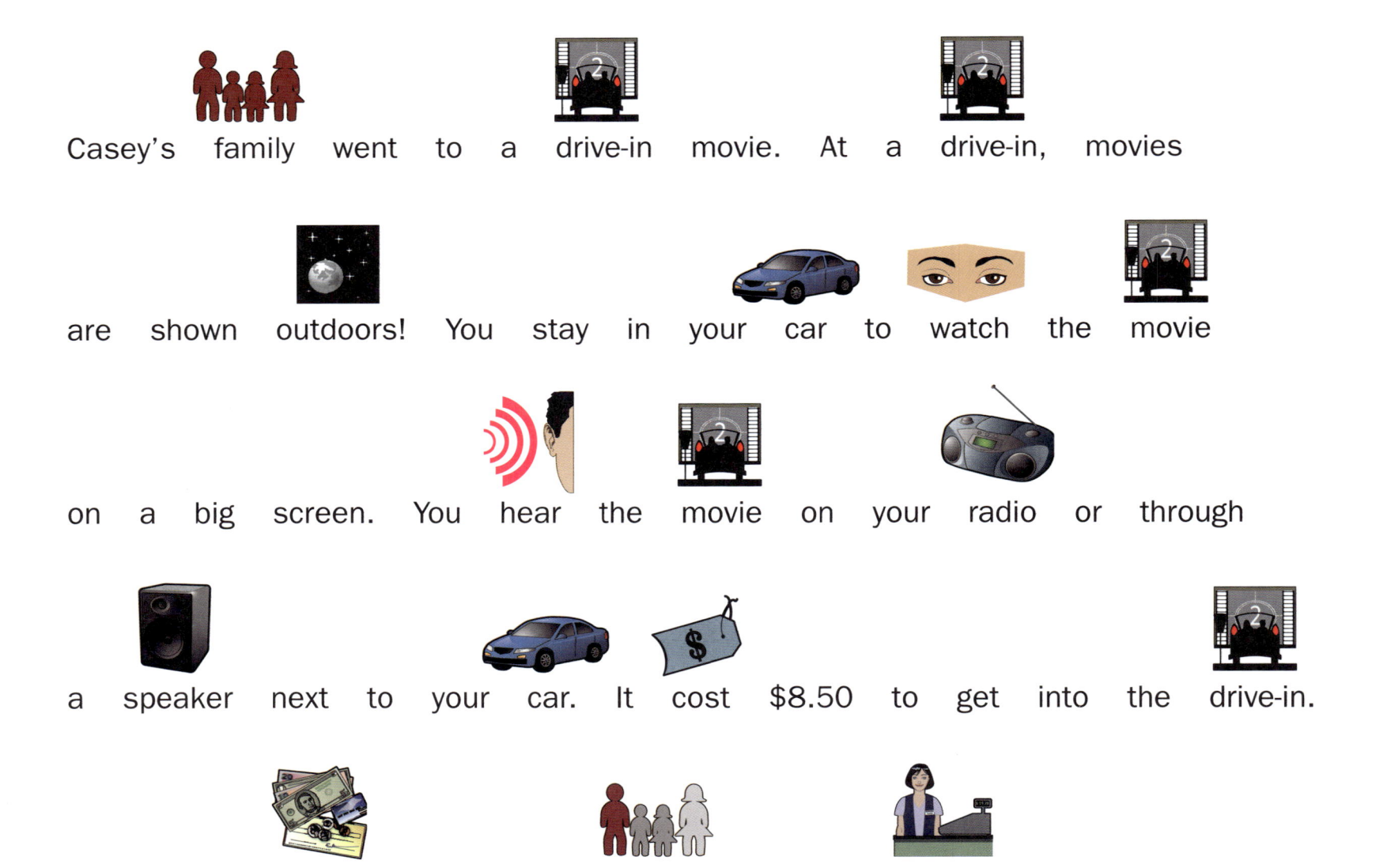

Casey's family went to a drive-in movie. At a drive-in, movies are shown outdoors! You stay in your car to watch the movie on a big screen. You hear the movie on your radio or through a speaker next to your car. It cost $8.50 to get into the drive-in. How much money did Casey's dad give the cashier?

WHAT DO WE NEED TO FIND OUT? CHECK THE BOX. ☑

☐ 1. How much money did Casey's dad give the cashier?

☐ 2. How far did Casey's dad drive to the movie?

1 2 3 4 5 6 7 8 9 10

Next dollar line

$ _____

Ordering lunch

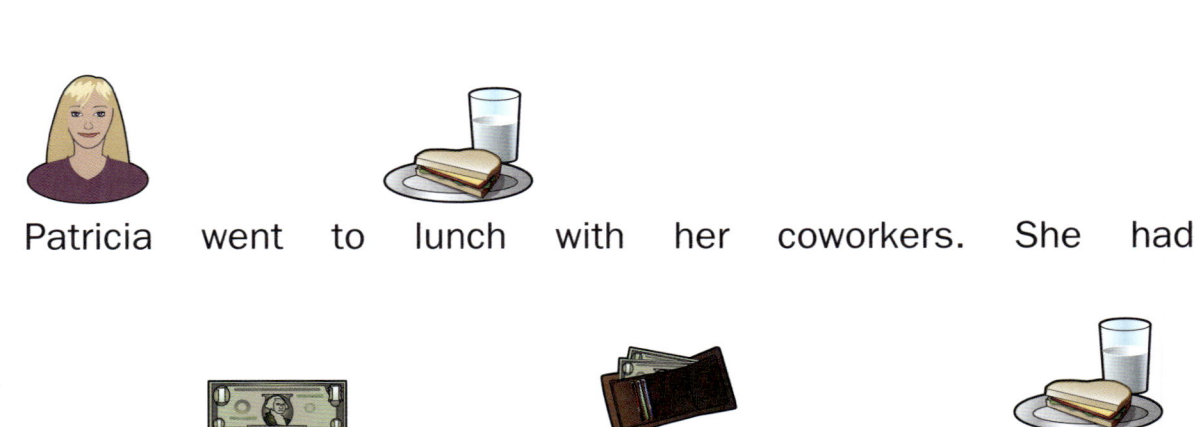

Patricia went to lunch with her coworkers. She had some dollar bills in her wallet to pay for lunch. She ordered the fruit and sandwich plate. Her bill was $5.89. How many dollars did Patricia give the cashier?

WHAT DO WE NEED TO FIND OUT? CHECK THE BOX. ☑

☐ 1. How many dollars did Patricia give the cashier?

☐ 2. What kind of fruit is in Patricia's salad?

1 2 3 4 5 6 7 8 9 10

Next dollar line

$ _____

Unit 4: Measurement

Out to dinner

Corrinna went out to dinner with her sisters.

She had some dollar bills in her wallet to pay for dinner.

She ordered the chicken salad plate. Her bill was $7.32.

How many dollars did Corrinna give the cashier?

WHAT DO WE NEED TO FIND OUT? CHECK THE BOX. ✓

☐ 1. How long did Corrinna wait for her dinner?

☐ 2. How many dollars did Corrinna give the cashier?

1　2　3　4　5　6　7　8　9　10

Next dollar line

$ _____

Unit 4: Measurement

Mark at lunch

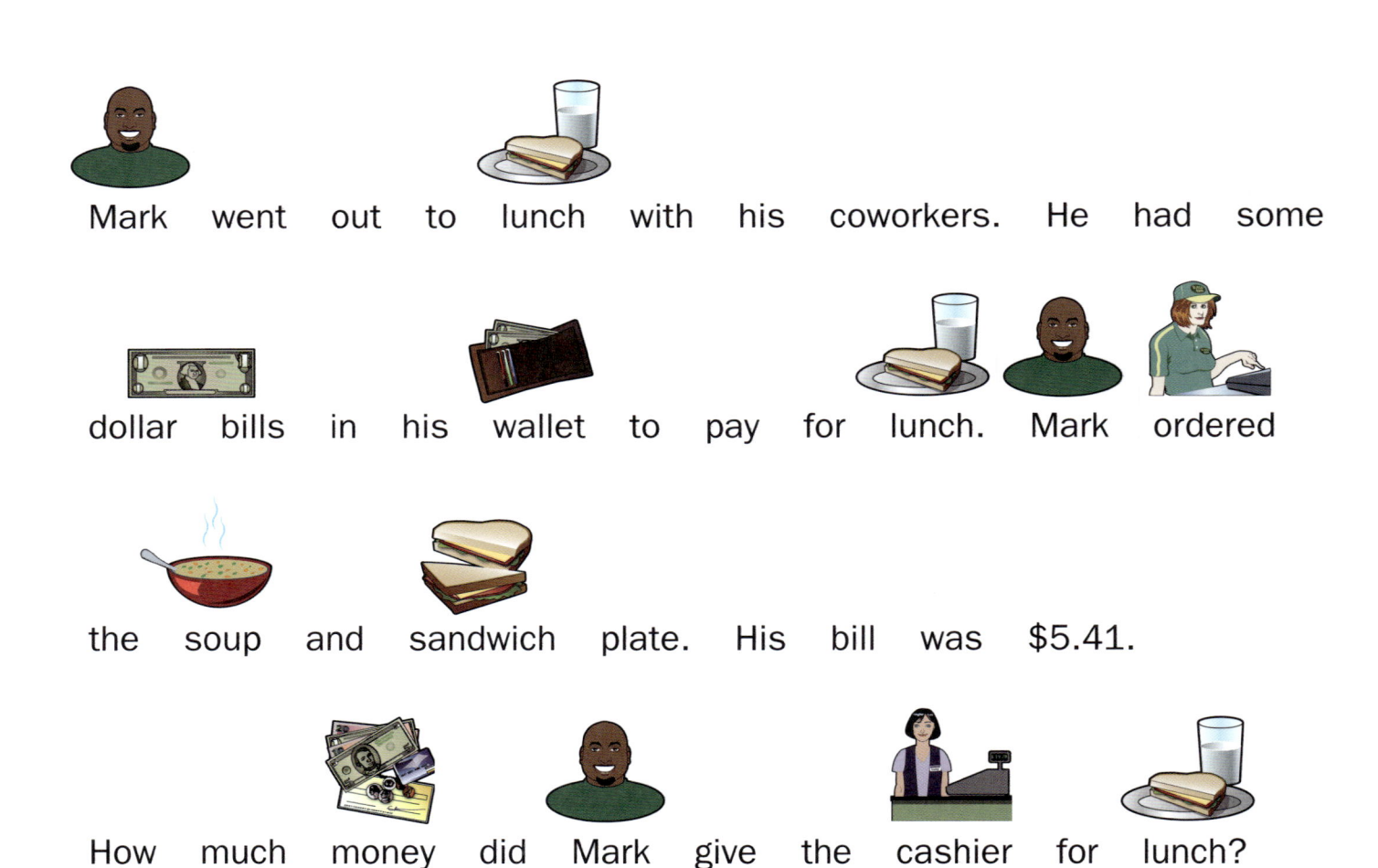

Mark went out to lunch with his coworkers. He had some dollar bills in his wallet to pay for lunch. Mark ordered the soup and sandwich plate. His bill was $5.41.

How much money did Mark give the cashier for lunch?

WHAT DO WE NEED TO FIND OUT? CHECK THE BOX. ☑

☐ 1. What kind of salad did Mark order?

☐ 2. How much money did Mark give the cashier for lunch?

| 1 | 2 | 3 | 4 | 5 | 6 | 7 | 8 | 9 | 10 |

Next dollar line

$ _____

Unit 4: Measurement

Kate's treat

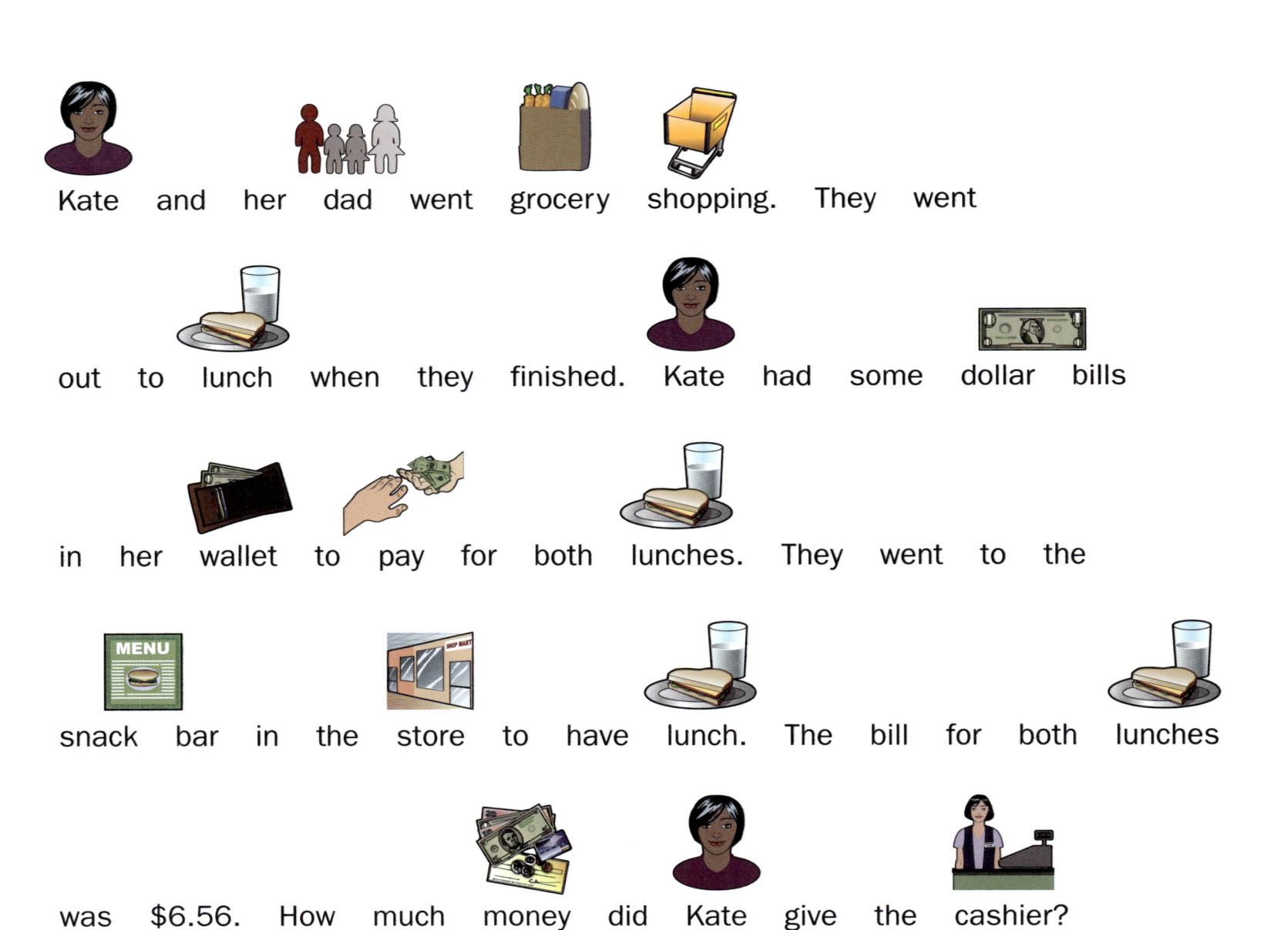

Kate and her dad went grocery shopping. They went out to lunch when they finished. Kate had some dollar bills in her wallet to pay for both lunches. They went to the snack bar in the store to have lunch. The bill for both lunches was $6.56. How much money did Kate give the cashier?

WHAT DO WE NEED TO FIND OUT? CHECK THE BOX. ☑

☐ 1. How much money did Kate give the cashier?

☐ 2. What item do Kate and her dad get next?

1 2 3 4 5 6 7 8 9 10

Next dollar line

$ _____

Unit 4: Measurement

Lunch at the Salad Ranch

Talya and Kia went to a restaurant called the Salad Ranch for lunch. Talya had dollar bills in her wallet to pay for lunch. She ordered the adult buffet, and Kia ordered the children's buffet. At a buffet, you choose foods from a large table. Altogether, the bill came to $7.49. How much money did Talya give the cashier?

WHAT DO WE NEED TO FIND OUT? CHECK THE BOX. ☑

☐ 1. How much money did Talya give the cashier?

☐ 2. How big is the buffet table?

1 2 3 4 5 6 7 8 9 10

Next dollar line

$ _____

Unit 4: Measurement

131

Shooting hoops

James loves to play basketball. He goes to the gym   to practice shooting hoops. Shooting hoops is another way    of saying making baskets. James bought a drink for $2.78 at the gym shop. How many dollars did James give the cashier?

WHAT DO WE NEED TO FIND OUT? CHECK THE BOX. ☑

☐ 1. How many dollars did James give the cashier?

☐ 2. How many baskets did James make?

| 1 2 3 4 5 6 7 8 9 10 |

Next dollar line

$ _____

Unit 4: Measurement

133

Movie day at the gym

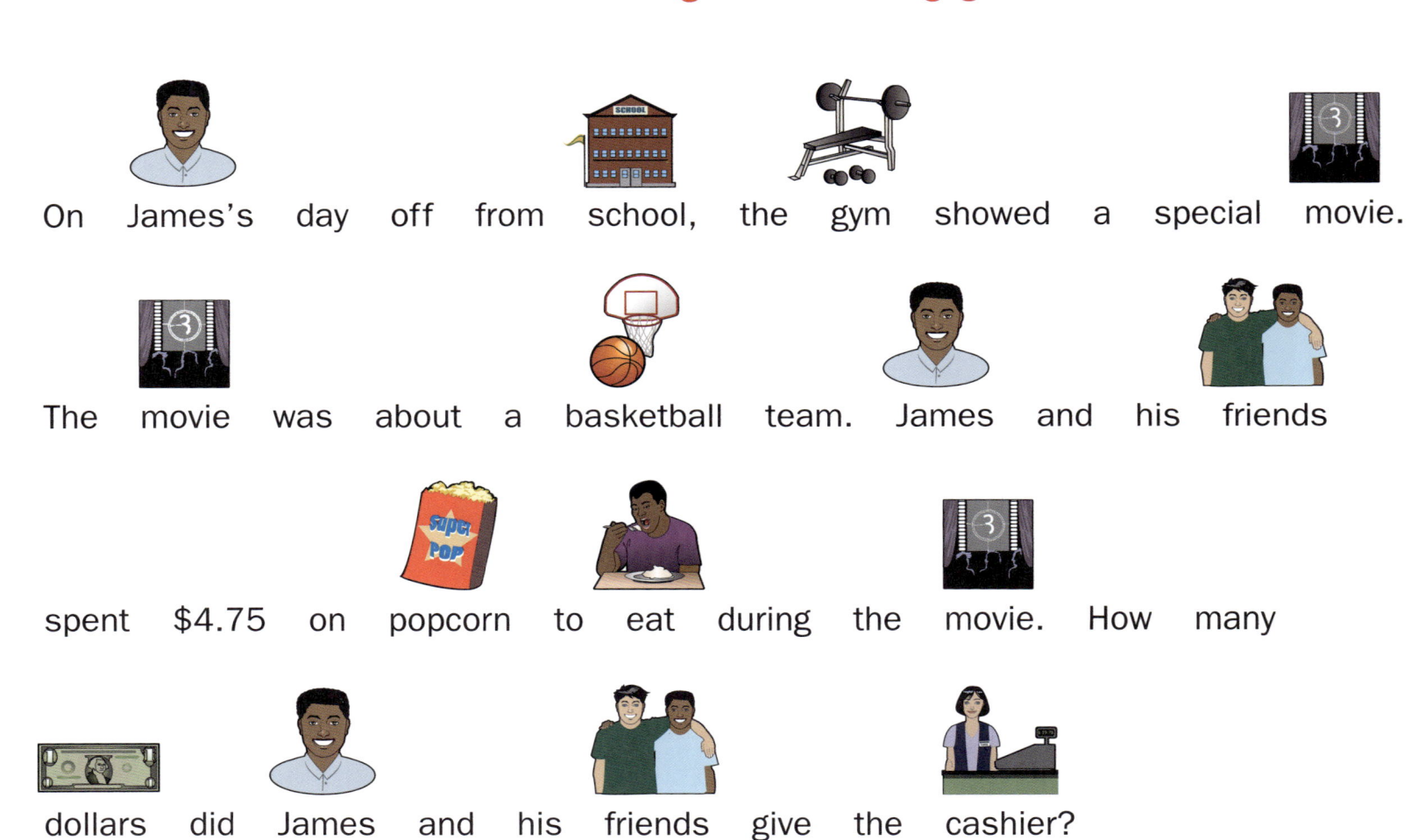

On James's day off from school, the gym showed a special movie. The movie was about a basketball team. James and his friends spent $4.75 on popcorn to eat during the movie. How many dollars did James and his friends give the cashier?

WHAT DO WE NEED TO FIND OUT? CHECK THE BOX. ✓

☐ 1. How many dollars did James and his friends give the cashier?

☐ 2. How many friends from the gym came to James's house?

1 2 3 4 5 6 7 8 9 10

Next dollar line

$ _____

Unit 4: Measurement

Lee's poster

Lee has a poster of his favorite band.

He wants to buy wood to make a frame   for his poster. The sides of his poster are 2 ft., 1 ft., 2 ft., and 1 ft.

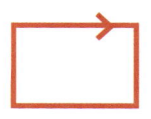

What is the perimeter of Lee's frame?

WHAT DO WE NEED TO FIND OUT? CHECK THE BOX. ☑

☐ 1. What is the perimeter of Lee's frame?

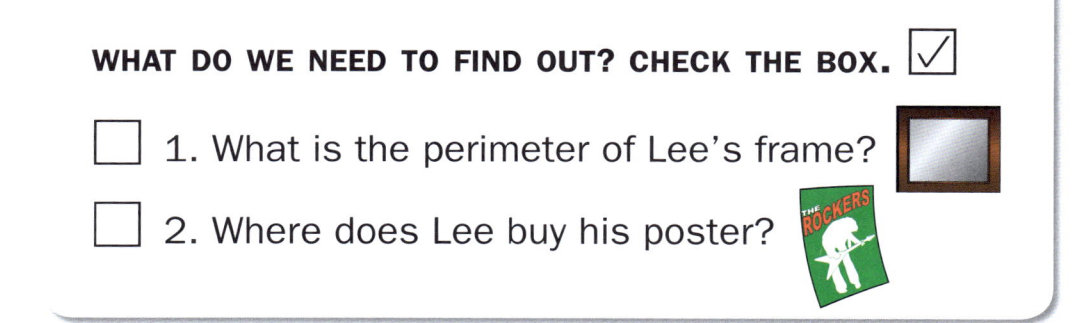

☐ 2. Where does Lee buy his poster?

Length	Length	Length	Length	Perimeter
☐ ft. +	☐ ft. +	☐ ft. +	☐ ft. =	☐ ft.

Perimeter measure

_____ ft.

Unit 4: Measurement

Carla's picture

Carla has a  picture of her  friends from  horseback riding. She wants to make a border with leather  string around the  frame. The sides of the frame are 1 ft., 2 ft., 1 ft., and 2 ft. What is the 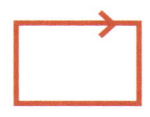 perimeter of  Carla's  frame?

WHAT DO WE NEED TO FIND OUT? CHECK THE BOX. ☑

☐ 1. How much does Carla pay for the picture?

☐ 2. What is the perimeter of Carla's frame?

| Length | Length | Length | Length | Perimeter |

☐ ft. + ☐ ft. + ☐ ft. + ☐ ft. = ☐ ft.

Perimeter measure

_____ ft.

Unit 4: Measurement

Carter and the rabbits

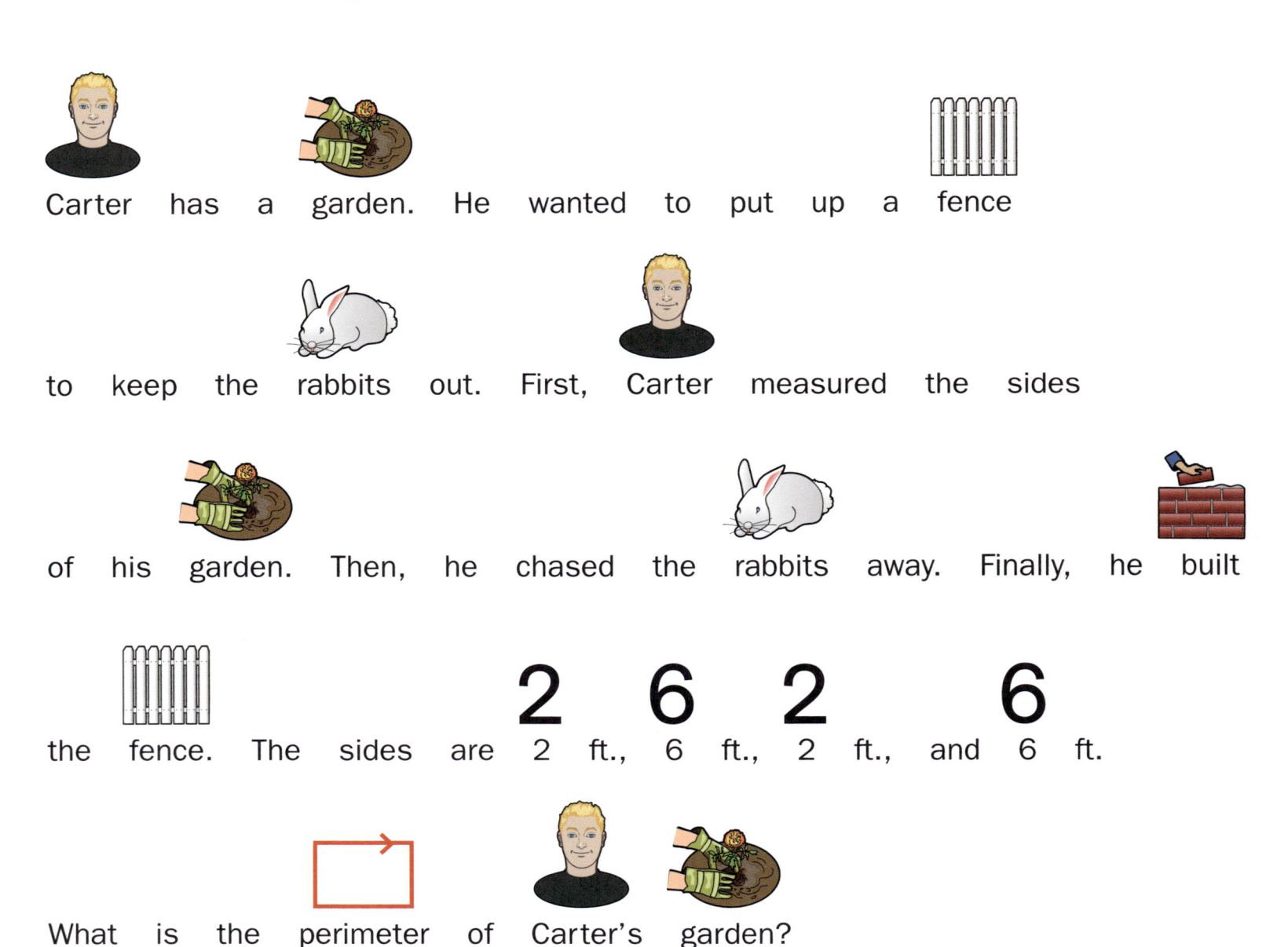

Carter has a garden. He wanted to put up a fence to keep the rabbits out. First, Carter measured the sides of his garden. Then, he chased the rabbits away. Finally, he built the fence. The sides are 2 ft., 6 ft., 2 ft., and 6 ft. What is the perimeter of Carter's garden?

WHAT DO WE NEED TO FIND OUT? CHECK THE BOX. ☑

☐ 1. Where did the rabbits come from?

☐ 2. What is the perimeter of Carter's garden?

Length	Length	Length	Length	Perimeter
⬜ ft. +	⬜ ft. +	⬜ ft. +	⬜ ft. =	⬜ ft.

Perimeter measure

_____ ft.

Unit 4: Measurement

141

Lee's wall

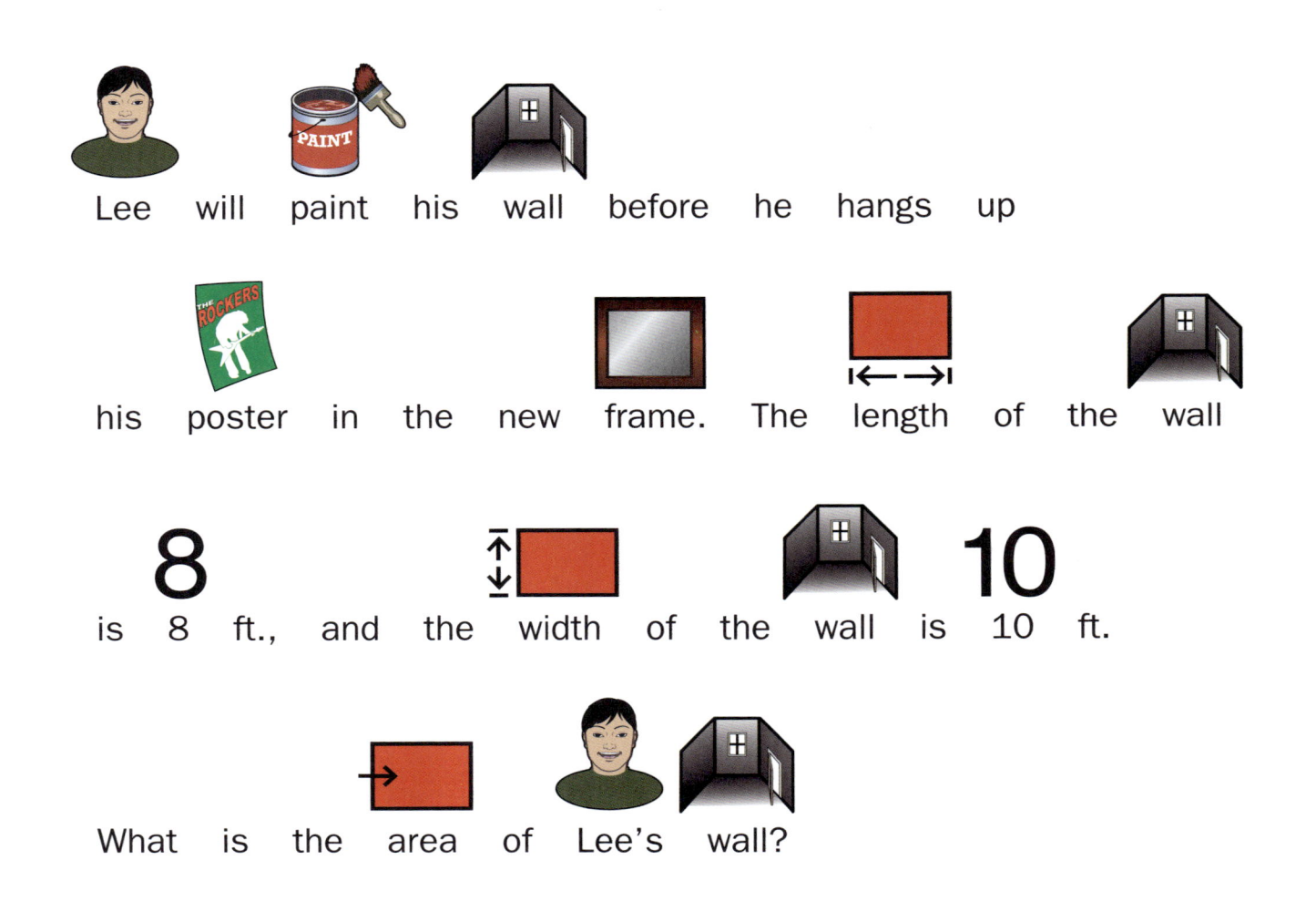

Lee will paint his wall before he hangs up his poster in the new frame. The length of the wall is 8 ft., and the width of the wall is 10 ft. What is the area of Lee's wall?

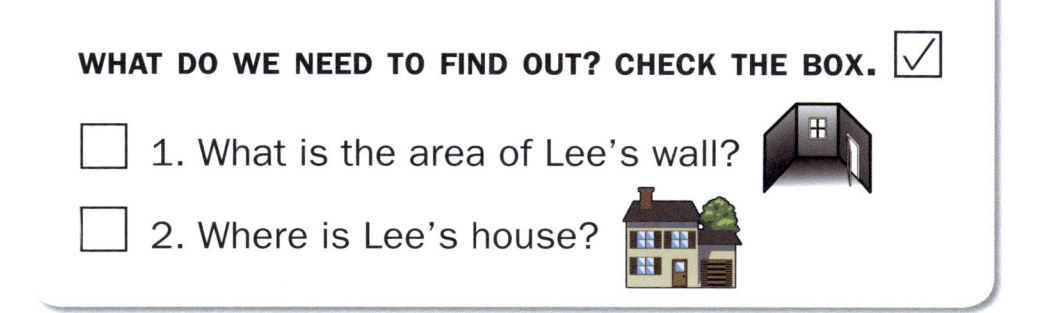

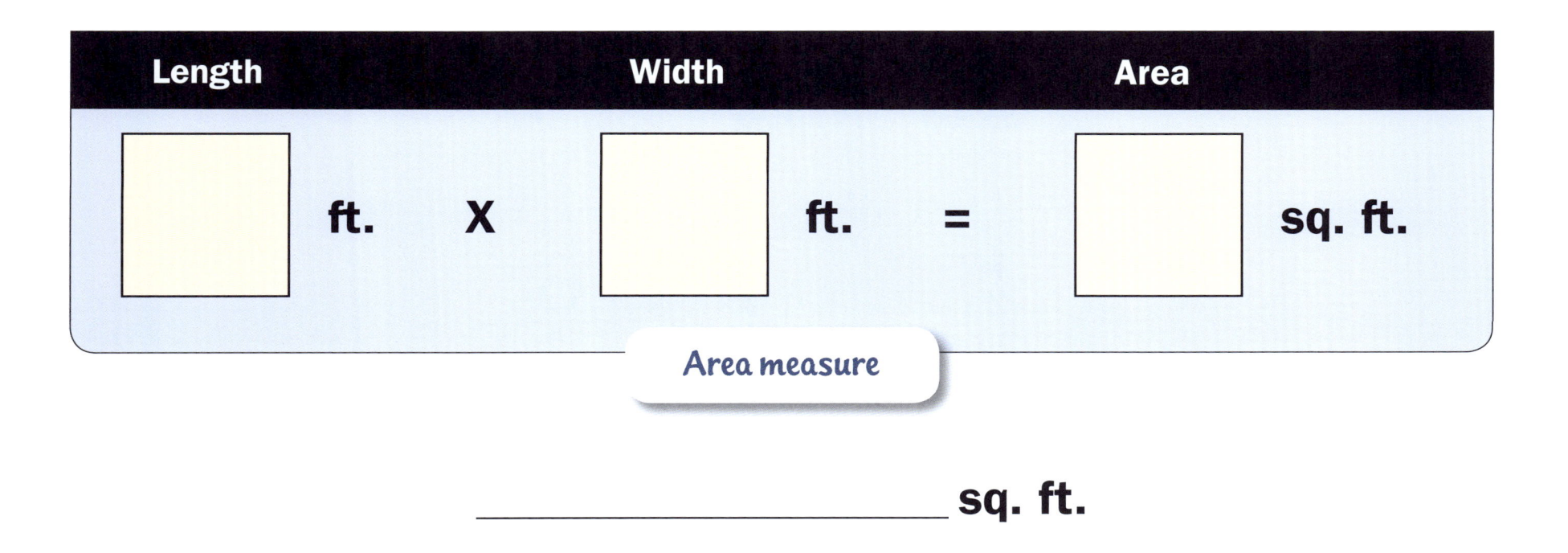

WHAT DO WE NEED TO FIND OUT? CHECK THE BOX. ✓

☐ 1. What is the area of Lee's wall?

☐ 2. Where is Lee's house?

Length	Width	Area
☐ ft.	X ☐ ft.	= ☐ sq. ft.

Area measure

_____ sq. ft.

Unit 4: Measurement

143

Carla's wall

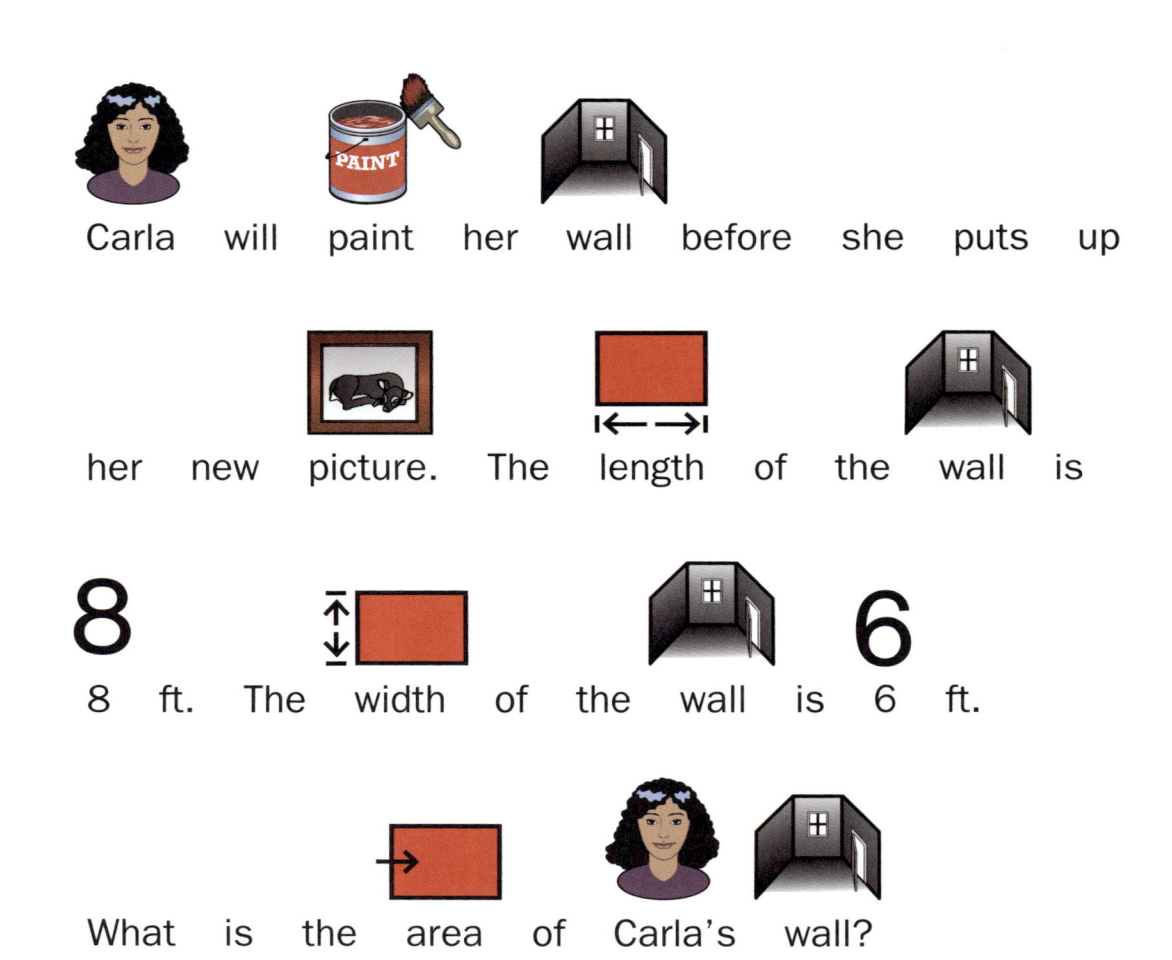

Carla will paint her wall before she puts up her new picture. The length of the wall is 8 ft. The width of the wall is 6 ft. What is the area of Carla's wall?

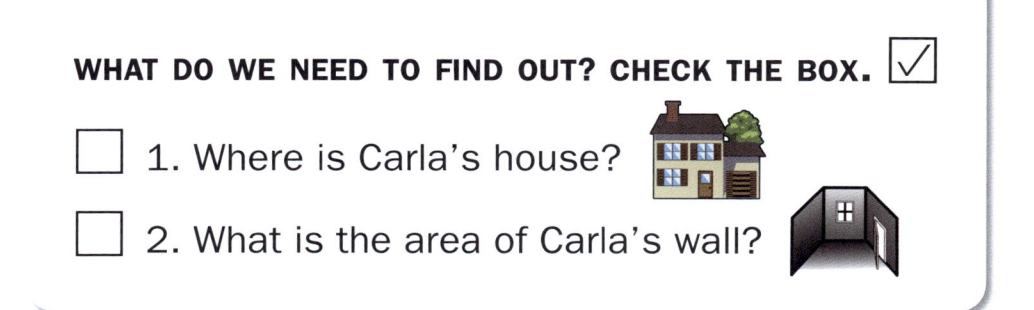

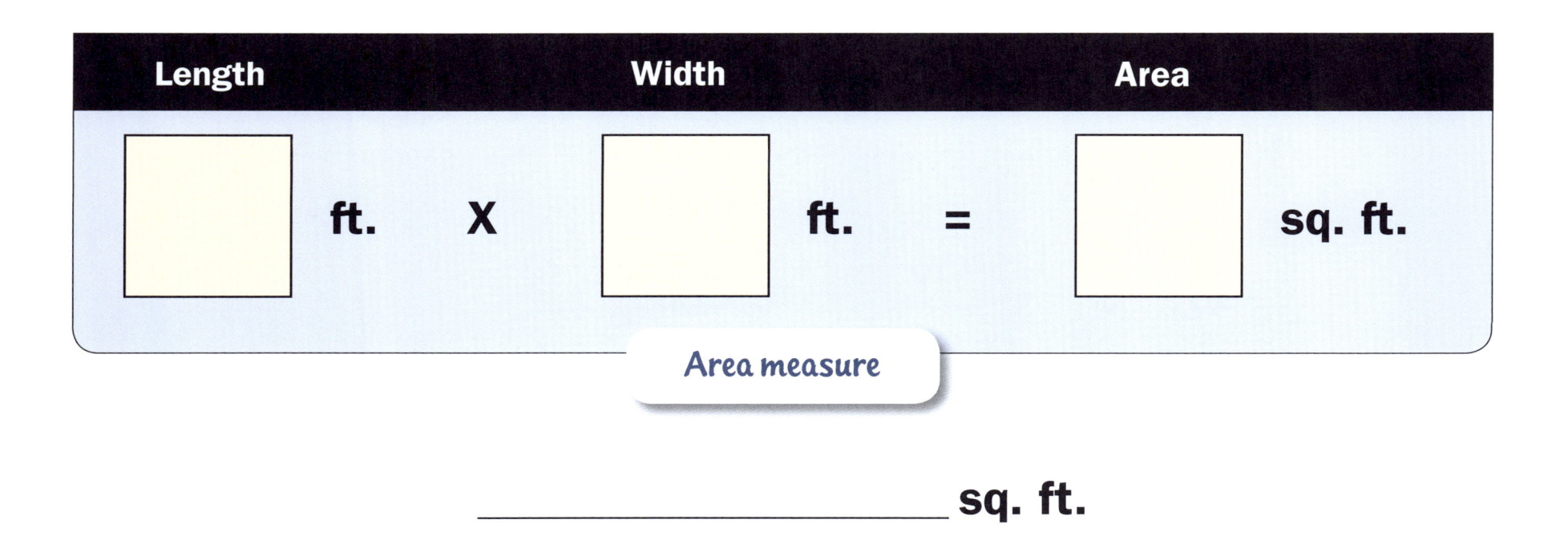

WHAT DO WE NEED TO FIND OUT? CHECK THE BOX. ☑

☐ 1. Where is Carla's house?

☐ 2. What is the area of Carla's wall?

Length	Width	Area
☐ ft. X	☐ ft. =	☐ sq. ft.

Area measure

_____ sq. ft.

Unit 4: Measurement

145

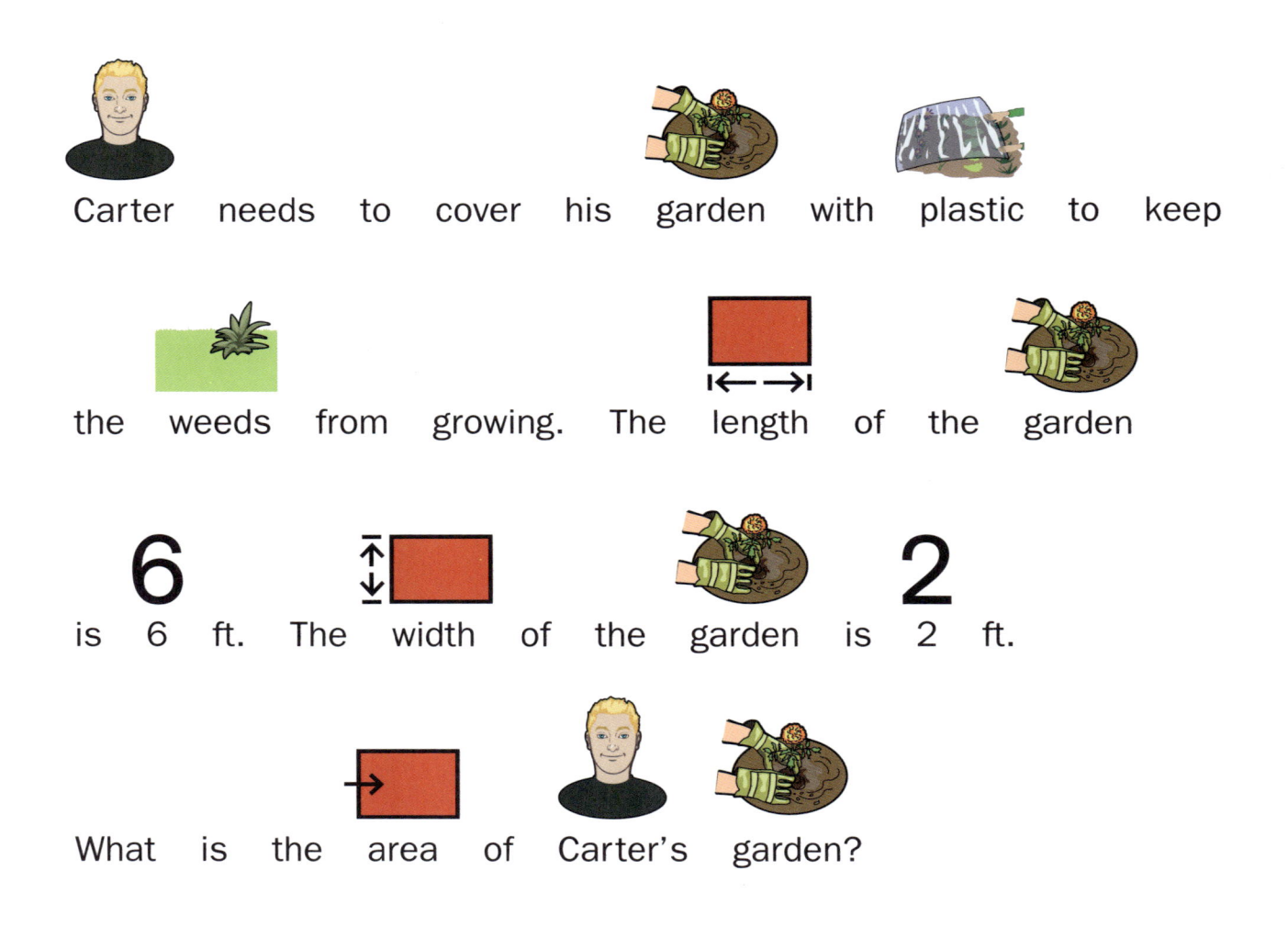

Carter's garden

Carter needs to cover his garden with plastic to keep the weeds from growing. The length of the garden is 6 ft. The width of the garden is 2 ft. What is the area of Carter's garden?

WHAT DO WE NEED TO FIND OUT? CHECK THE BOX. ☑

☐ 1. What is the area of Carter's garden?

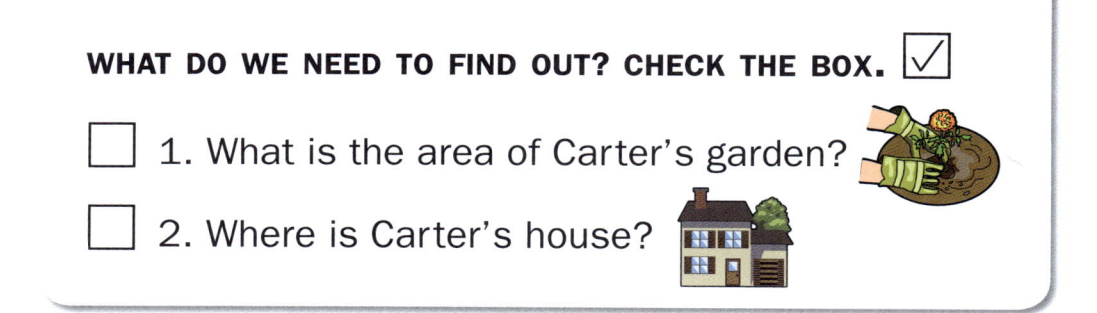

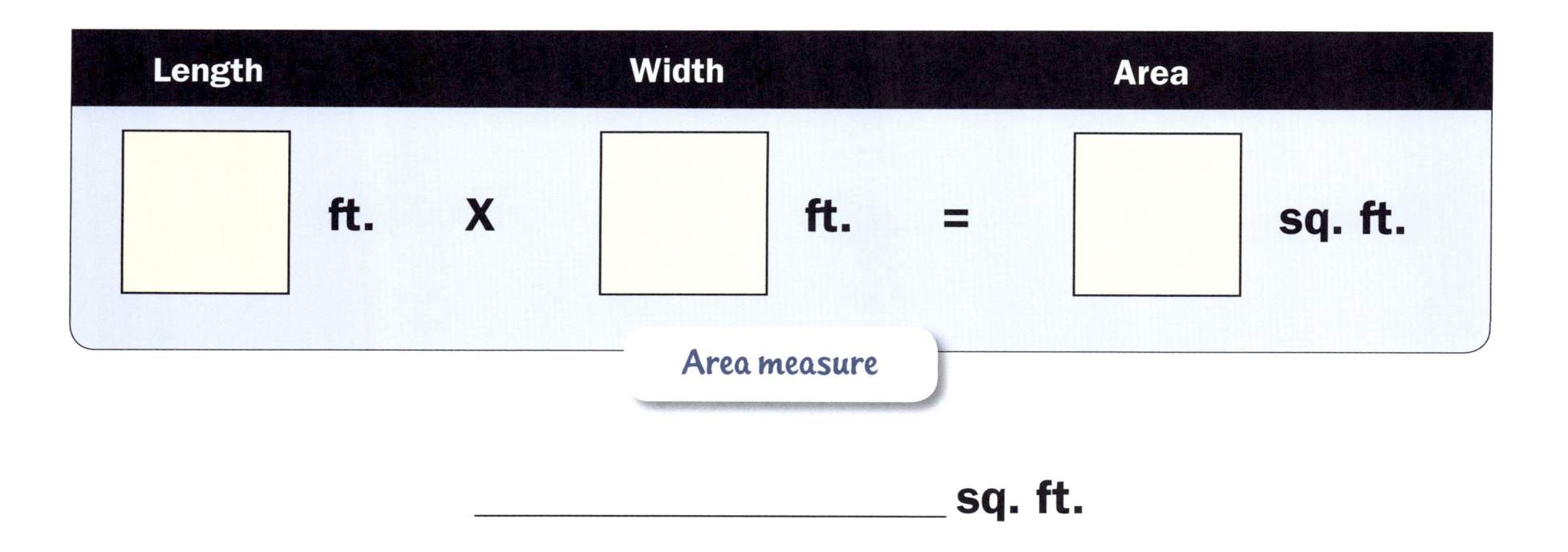

☐ 2. Where is Carter's house?

Length	Width	Area
☐ ft.	X ☐ ft.	= ☐ sq. ft.

Area measure

_____ sq. ft.

Unit 4: Measurement

147

A lunch date

Caleb and Ella went out for a lunch date. They ordered vegetable soup and a turkey sandwich. The bill was $7.75. How much money did Ella give the cashier?

WHAT DO WE NEED TO FIND OUT? CHECK THE BOX. ✓

☐ 1. Where did Ella go after lunch?

☐ 2. How much money did Ella give the cashier?

1 2 3 4 5 6 7 8 9 10

Next dollar line

$ _____

Unit 4: Measurement